AKNOWLEDGEMENTS:

To my wife Cristina, to my sons Ricardo and Miguel and to my parents.

To all of my colleagues of 6 countries who gave me their own communication stories.

To Jorge Conceição, a careful reader of my books who suggested that I write about this topic.

To Edgar Potter for his translation work.

To my publisher, Edições Silabo, and especially to my editor, Manuel Robalo, whose is suggestions greatly improved this book.

COMMUNICATE 2.0

The Art of Communicating in the 21st Century

Filipe Carrera

EDIÇÕES SÍLABO

It is forbidden to reproduce in whole or in part, under any form
or by any means, **including photocopying**, this work.
The transgressions will be liable to the penalties provided in legislation.

Sílabo can be visited in:

www.silabo.pt

Editor: Manuel Robalo

FICHA TÉCNICA:

Title: Communicate 2.0 – The Art of Communicating in the 21st Century
Author: Filipe Carrera
© Edições Sílabo, Lda.
Cover: Pedro Mota
Translation: Edgar Potter

1st Edition
Lisbon, 2012.
Printing: Publidisa
DL: 343023/12
ISBN: 978-972-618-681-6

EDIÇÕES SÍLABO, LDA.
R. Cidade de Manchester, 2
1170-100 LISBOA
Telf.: 218130345
Fax: 218166719
e-mail: silabo@silabo.pt
www.silabo.pt

Table of Contents

Preface   17

Introduction   19

Chapter 1
COMMUNICATE, WHAT FOR?

The Essence of Communication   23
- Framework   23
- Who with?   23
- What?   23
- Medium?   24
- When?   25
- Why?   25
- The feedback   26

Multimedia Communication   27
- Framework   27
- Choosing the most effective medium   27
- Communication objective   28
- Communication methods   28
- Face-to-face communication   28
- Multimedia communication   29
- Oral communication   29
- Written communication   29

Chapter 2

FACE-TO-FACE COMMUNICATION

Preparation	33
Framework	33
Communicating is transmitting emotions	33
What is my style?	34
The role of the organization	34
Check list	35
Having a plan	35
Plan format	35
Mind map	36
Formats	38
Framework	38
Training	38
Trainer as catalyst	38
Learning styles	39
Sales	40
Structure of sales presentation	40
Project proposal	41
Project report	41
Briefing	41
Briefing structure	42
Speech	42
Structure of a speech	43
«And now for something completely different»	43
TED is unstoppable	44
Debate	45
Debating well...	45
Audience Analysis	46
Framework	46
How to analyze the audience beforehand	46
And when there is no prior analysis of audience	47

Aspects to consider in the analysis	48
Demographics	48
Knowledge	48
Composition	48
Expectations	49
Origin	49
International Audiences	50
Framework	50
Preparation	50
Characterizing an international audience	51
Homogenous audience	51
Heterogeneous audience	51
Heterogeneous audience with predominance	52
Linguistic barriers	52
Interpretation	53
Precautions with interpretation	53
Cultural aspects	54
Tuning	55
Feedback	56
Gestures	57
Sign curiosities	57
Some additional precautions	58
Structure of the Presentation	59
Framework	59
Structure-type	59
Receiving participants	59
Opening	60
Breaking the ice	60
Content	61
Closing	62
Follow-up	62

Verbal Communication	63
Framework	63
«All brain» approach	64
Speaking to the senses	64
Telling a story	66
Precautions with stories	66
The power of the word	67
Beginning	67
Good starts	68
Apologizing	68
This is good, but...	68
Non-Verbal Communication	69
Framework	69
The subconscious at the wheel	69
Connection between sub-consciousnesses	70
Posture	70
Gestures	71
Tone of voice	72
The power of silence	73
Eye contact	74
Controlling non-verbal communication	75
The Room	76
Framework	76
Room analysis	77
Set-up styles	78
Theater	78
Auditorium	79
Banquet	80
Classroom	81
Meeting room	82
U-shape	83
Standing	84

Chairs	85
Support rooms	85
Resources	**86**
Framework	86
Choosing resources	86
Audio resources	86
Visual resources	88
Overhead projector	88
Flipchart	89
Advice for using a flipchart	90
PowerPoint	91
Advice for using PowerPoint	92
Non-linear presentations	93
Prezi	93
Mind map	94
Support Equipment	**95**
Framework	95
Graphics tablet	95
Wireless presenter	97
Video projector	98
Advice for using video projectors	99
Electronic voting equipment	100
Dealing with Difficult Situations	**101**
Framework	101
Difficult participants	102
Equipment malfunction	102
Problems with files	104
Unexpected audiences	105
Almost empty room	105

Chapter 3

MULTIMEDIA COMMUNICATION

Collaborative Platforms	109
Framework	109
What are they?	109
Features	109
Preparation	110
During	111
Afterwards	111
Practical applications	111
Skype	112
Framework	112
What is it?	112
One way to use it	113
Advantages	114
Practical advice	114
Are chats just idle chatter?	115
The mystery of the video call	116
Videoconference	116
Framework	116
Mental blocks	117
Range of coverage	117
Preparation	118
During	119
Afterwards	120
Applications	120
Vodcast and Videos	121
Framework	121
The multimedia world	121
Mental barriers	122
Equipment	123
Microphone	123

Software	125
Creation process	125
Script creation	126
Choice of location and wardrobe	126
Video recording	126
Editing the video	128
Final result	128
Advice and precautions	129
Applications	131
A 2.0 example	131
Live	133
Television Programs	134
Framework	134
Preparation	134
During	135
Afterwards	135

Chapter 4
AUDIO COMMUNICATION

Telephone	139
Framework	139
Preparation	139
During	140
Afterwards	140
Applications	140
Calling for free?	141
Conference Call	142
Framework	142
Conference call types	142
Preparation	142
During	143
Afterwards	144

Applications	144
And the costs	145
Podcast	147
Framework	147
What is a podcast?	147
Making a podcast	147
Equipment	148
Software	149
Creation process	149
Creating the script	149
Recording audio content	149
Editing audio	150
Final result	150
Advice and precautions	150
Applications	151
Radio	152
Framework	152
Preparation	152
During	153
Afterwards	153

Chapter 5

WRITTEN COMMUNICATION

E-mail	157
Framework	157
80% of e-mail is useless	157
The e-mail in Aramaic	157
Forget grammar school rules	158
A conversation is worth more than a thousand e-mails	159
The toilet paper type of e-mail	159
Here goes a cc for all	160

The 13th version e-mail	160
The e-mail confirming the receipt of an e-mail	161
Scheduling meetings through an avalanche of e-mails	161
The useless information e-mail	162
How cool is my smartphone?!	163
Very polite, not very assertive	164
All e-mails will be forwarded	165
Reports and Proposals	165
Framework	165
Most common mistakes	166
Preparation	166
Organization of topics	166
Writing	167
Principles of the Technical Writing method	167
Articles	169
Framework	169
Prior questions	169
Preparation	169
Practical advice	170
There is life after publishing	170

Chapter 6
PRACTICAL ADVICE

Preparation	175
Framework	175
Promotion	175
Time management	175
Visualization	176
Self-motivation	176
Checklist	177

During	178
Framework	178
Empathy	178
Questions	178
Interactivity	179
Humor	179
Multitasking	179
Feedback	180
Framework	180
Feedback phases	180
Receive	180
Appreciate	180
Summarize	180
Ask	181
Example of online feedback	181
Going Further	182
Framework	182
Seeing the best	182
Volunteering	182
AIESEC	183
JCI	184
Toastmasters	185

Chapter 7

CHALLENGES AND SOLUTIONS

Communicators of the World, Unite!	189
Framework	189
Long distance intercultural communication	189
Use of collaborative platforms	190
When our nerves take over	191
Being omnipresent	192
Communicating on Facebook	193

And what if everything goes wrong?	195
Get people communicating	195
The element of surprise	196
Overcoming cultural barriers	197
Dressed to communicate	198
Going further	199
Verbal communication	199

Bibliography 201

Preface

It is happening all the time and is literally everywhere. How, what and when we say something can have consequences to our personal and business lives that are not immediately obvious. More over we can never under estimate the impact of this on those we are communicating with.

What we now recognize as communication has greatly altered in the past two decades. Media communication was once the exclusive preserve of media corporations. This is no longer the case with the internet and the development of sites like You tube, Facebook and Twitter.

Resources I frequently employed when I was JCI President in 2008 and when I wanted to instantly reach hundreds of thousands of members scattered around the globe.

Now we can all «Broadcast», now we are all in some way or another part of the «Media», through «posts», «Likes», «Comments» and «Tweets». Now we all have an audience. No longer is it confined to direct one to one communication or to whom we directly address, be it an audience of one or one hundred.

Yet the traditional modes of communicating, the impact of that one to one meeting, the eye to eye contact has never been more important. Perhaps the development of the electronic forms of communication has in fact heightened the value of those more traditional and for many, familiar methods.

This books deals with both and Filipe Carrera is well placed to do so. I have had the privilege to work with Filipe on different projects over the last three years in different parts of the world. He knows well the value of communication both with the traditional methods and the more modern electronic methods. He has measured and studied the impact of both and has developed many useful training sessions to help people understand how best to use them to further their personal or business goals.

Together we have trained audiences internationally in different parts of the world and his experience in understanding and recognizing the importance of cultural differences in this area is second to none.

I am sure that no matter what level of communicating you are doing or what level of communication skills you believe you have, you will find this book an excellent resource and tool for your own personal development or perhaps to further you professionally.

The fact is you are no different to anyone else. You are communicating, so you should read this book.

Graham Hanlon
JCI President 2008.

Introduction

The goal of this book is to help the reader overcome the greatest fear of any human being: speaking in public.

I myself was a victim of this fear. Up until my third year in university I was totally incapable of speaking in public, no matter how small the audience. But everything changed; I was the one that made the change, because no one can do it for us.

I am aware that in the world in which we live, our quality of life depends on how we communicate. This is a skill that can be perfected over a lifetime; there are persons who are good communicators by nature and there are others who are terrible communicators until they finally discover their communicator rib and change the world.

My intention is to share with the reader a set of techniques and experiences that I have been gathering in my almost 20 years of a professional career that has involved a lot of communication.

Besides being a professor at various universities for several years, I have built up a professional career as a trainer and international speaker in nearly 50 countries on 4 continents, before widely different audiences.

Note: Communicating is not just speaking; it's motivating others to make a positive change, so that we have to be careful with our verbal and non-verbal languages, even if this means eliminating some mental programs that limit us.

I have been through many situations and I have used widely diversified means to communicate, specifically resorting to the web. I think that communication in the 21st century has to change according to the technologies accessible to us and that allow us to communicate with the world.

Furthermore, I believe that professionals who do not embrace these new ways of communicating may face the same type of painful problems their predecessors faced 20 years ago with the computerization of businesses.

For this reason, I have chosen to adopt an approach that embraces the various forms of communication that are at the disposal of any professional in the 21st century, structuring the book so that it can serve as a handbook for quick consultation in any communicational situation.

So I hope, dear reader, that the following chapters are very useful to you as a communicator, and consequently as an agent of change.

1

COMMUNICATE, WHAT FOR?

The Essence of Communication

Framework

We all communicate on a daily basis, or at least we think we do, but even so we invest very little time analyzing the way in which we communicate, even though it is at the root of our successes and failures, whether they be professional or personal.

Before we start a presentation, report, phone call, video conference, or any other form of communication, we should keep in mind the following questions:
- Who are we trying to reach?
- What do we want to communicate?
- What medium will we use?
- When will we do it?
- Why will we do it?
- How will we guarantee feedback?

Who with?

We occasionally meet somebody with an innate ability to communicate in any circumstance, and we think that it must be someone who is gifted with an intrinsic quality, more or less mysterious, that we sometimes call charisma.

But, when we start to analyze the situation more carefully, we see that it is the audience who are the basis for interesting communication, in other words, the first question should be: «Who are we communicating with?» This, in turn, is subdivided into a series of further questions:
- What are their needs?
- Expectations?
- Experience?
- Level of education?
- Etc.

What?

Over the course of our academic and working years, we are led to believe that the content of our message is the most important element of communication, with the method playing a somewhat less important role.

I'm not saying that we can communicate any foolishness, with positive results, just because we do so correctly. What I want to make clear is that we can have perfect content, but if we communicate it poorly we will certainly not achieve the intended results.

Curiously, bad content communicated correctly gives us a better chance of reaching our objectives, which explains the initial success of many totalitarian regimes.

Medium?

Is it the same to communicate over the phone as in an auditorium? Certainly not.

We must adapt our communication to the chosen medium, taking into consideration the level of non-verbal communication that the medium allows.

Level of Non-Verbal Communication	Media
High	• Presentation in person • Television program • Video conference • Webcast • Online videos
Average	• Teleconference • Radio program • Podcasting
Low	• E-mail • Report

Non-verbal communication is very effective in communicating complex emotions; for example, it is extremely difficult to write a truly motivating e-mail; any other medium with a high level of non-verbal communication would be more effective.

Therefore, when we want to motivate, manage conflict or create a team, we should always use media with a high level of non-verbal communication. The media with lower levels should be used to supplement.

The medium may be chosen by us or given from the start; but in any circumstance it should be taken into consideration, keeping in mind that, regardless of the medium, we can always increase or reduce its level of non-verbal communication depending on the results we wish to achieve.

When?

The moment we communicate can be a very relevant factor, there being two principal possibilities:

- The moment is clearly established in time: there is a day and an hour in which we find ourselves with our recipients, for example, in a presentation, a teleconference, or a live radio program.
- The moment is not established: The recipients will be exposed to the communication at a time defined by themselves or a third party, for example a video, a recorded television program or an e-mail.

It is preferable to define the moment in which the recipients will receive our communication, since we can choose it depending on previous communications, biological clocks, etc.

If the moment is defined by a third party, we must keep in mind the time of day in which we will communicate, as we may be facing an audience that is fresh and prepared to absorb everything or, on the other hand, our target audience may be tired after a day at work or a busy social agenda from the previous day.

Likewise, our communication may be one of several in a program, which makes it imperative to analyze the communications before and after ours to ensure that we are well contextualized.

If, on the other hand, the moment is not defined, we should avoid mentioning anything that would refer to when our communication took place, unless it is essential to clarify that certain information is not available at the time, which may later be known by the recipients as they receive our communication.

Why?

If we have a well-defined objective in line with our audience's needs and expectations, we will have a better chance of success in any type of presentation.

Why do people invest time and/or money to listen to us? What is the value that we provide? Or, if you like, what would be the return on investment? Nobody minds investing, but we do not like wasting time and/or money on something that does not meet our needs.

Therefore, determining the why starts by analyzing our recipients and what spurs them to action.

The feedback

There is no communication without feedback; therefore we must establish *a priori* how we are going to obtain that feedback.

In a presentation in person, that task may be easy, as we can analyze the non-verbal communication in the room (yawns, nods, smiles...) or even question the participants directly.

Other forms of communication (podcast, TV program...) may be more difficult; however, thanks to online social networks, this task has become much easier.

Furthermore, even when we do a presentation in person, we can better measure our impact through social networks. As I am a compulsive user of social networks, I have gotten used to verifying the feedback of any communication method on these media in the following ways:
- Mentions on Twitter.
- Comments and pictures shared on Facebook.
- Videos shared on various platforms.
- Articles on blogs, communication media and other sites.

I would say that feedback is a continual process in three phases:

- **Before communication**: in the process of determining the needs, it is extremely important to have the feedback of the organizers and the potential participants.
- **During communication**: if interaction with the audience is possible, we may gather feedback through the audience's non-verbal communication or by directly questioning a few participants.
- **After communication:** using traditional evaluation questionnaires, speaking with participants or even observing reactions on social networks, the web in general and in other communication media.

The advantage of planning from the beginning how we are going to gather the participants' feedback is that it allows us to gain a clear vision of the data gathered and to reach conclusions for on-going improvement.

Multimedia Communication

Framework

Let's do a small visualization exercise: imagine a good communicator. I bet the image that came to mind was of somebody in a prominent location in a room, probably using a microphone.

The definition of a good communicator is intimately linked to our idea of face-to-face communication or of a television program, but we are in the 21st century and if there is anything that characterizes the last 10 years it's the democratization of communication, its broader reach and greater variety.

If we want to be successful professionals in this new reality, we must go beyond the old techniques of presentations and learn to communicate over multiple media.

Choosing the most effective medium

Choosing the most effective medium for communication will be a function of the following variables:

- The intended audience's presence and profile – we may have the best content and use the best communication techniques, but if we do not have recipients for that communication, then it will be a wasted effort.
- Communication objective – this decidedly influences the medium to choose, in conjunction with the remaining variables.
- The communicator's skills – we all know cases of communicators who are wonderful in one medium, for example, over the phone, but horrible in another medium, for example when holding a meeting in person. However, because we're talking about skills, this is a component capable of being continually improved; there is

always hope for those who are currently bad communicators, who at any and every opportunity may go on learning.
- Logistical questions – coming into play in this category are issues such as:
  — Time for preparation and execution.
  — Available technical means.
  — Budget.
  — Size of target group.
  — Etc.

Communication objective	Regardless of the medium, in professional terms we communicate with the following objectives: • Propose – we present a proposed action, with the objective of gaining approval. • Sell – we seek to sell a product, service, idea or concept. • Report – we create a status report on a project, with the objective of informing the audience. • Educate – this is a more complex objective, as it seeks to make a positive change in the audience's skills, knowledge, and motivations.
Communication methods	Because I believe that in this world we need to have the skill set to communicate through various interconnected methods, I will deal with several methods in this book, divided into the following categories. • Face-to-face communication. • Multimedia communication. • Oral communication. • Written communication.
Face-to-face communication	As face-to-face communication, we will consider the following forms: • Training initiatives. • Sales. • Report presentations. • Briefings.

- Speeches.
- Debates.

As multimedia communication, we will cover the following:
- Collaborative platforms.
- Skype.
- Video conferences.
- Online videos.
- Television.

Multimedia communication

As oral communication, we will study the following:
- Telephone.
- Teleconference and audioconference.
- Podcast.
- Radio.

Oral communication

Finally, as written communication, we will cover the following types:
- E-mails.
- Reports and proposals.
- Articles.

Written communication

2

FACE-TO-FACE COMMUNICATION

Preparation

Framework

Face-to-face communication is without a doubt the most complex, due to the level of detail in the established communication.

We merely need to consider that less than 10% of communication is through what we say; the way in which we say it is the most important factor. In other words, as human beings we are very used to communicating emotions.

Communicating is transmitting emotions

I will give you an example based on your own experience. Remember an excellent presentation, where you left the room feeling very good, but somebody asked you questions regarding the exact content of the presentation, and you probably had a hard time recalling exactly what was said, possibly making comments similar to these:
- «It was an inspiring session».
- «A charismatic public speaker».
- «The presenter is a true communicator».
- «I felt like he was speaking to me».
- «You could tell the speaker is passionate about the subject».

These comments do not directly refer to the content, but they explain feelings and emotions that were transferred by the presenter to the audience. The interesting thing is that, should we ask the presenter to explain the success of his presentation, he would say the following:
- «I was well prepared».
- «There was a positive energy in the room».
- «The participants were receptive».
- «The audience was participative».

Before we even prepare any communication, we must be conscious of the importance that the method holds; therefore we should be attentive to a vast set of details that surround the act of communicating.

What is my style?

I have helped many people all over the world improve their communication skills, and any time I start training in this area I say the same thing: «Don't imitate; create your own style of communication».

Except for good actors, the majority of people cannot coherently imitate a style of communication and, if they did, they would come off as sounding fake and lose all credibility with the audience.

But if I don't have experience in giving presentations, how can I have a style? The answer is simple: «You learn as you go!»

I usually compare communication to driving: when we get our driving license and start to drive alone, our hands firmly grasp the steering wheel and our head moves in a strange way since we have not yet mastered our peripheral vision.

After a few thousand kilometers, we begin to relax and discover we are more capable than we thought, taking on our own driving style. Practice and time help us to become more comfortable and confident.

Our presentation style will be important in the preparation phase, as it is from there that we can determine what conditions are necessary for us to be successful in our communication.

The role of the organization

Whoever organizes small meetings or large conferences has an extremely important role in the success or failure of these events; however, I don't think this role is given its due importance, sometimes by the speakers themselves, who place themselves on a pedestal.

I have always tried to establish a very frank and open dialogue with the various organizations with which I have worked, not only because I recognize the importance of true teamwork between the organization and the presenter, but also because I believe the organization has vital information to help my presentation be a success.

It is equally important to know who does what, before, during and after an event so that there aren't any gaps or overlaps.

Check list

There is a series of questions that we should ask the organization before any event; I list here the ones that I see as essential:
- What is the event's objective?
- Who are the participants and how many will there be?
- Who is the organizer and/or sponsor?
- Where will it be held?
- How much time is allotted for the presentation?
- Who are the other speakers?
- What equipment will be available?
- What type of event will this be: formal or informal?

Having a plan

There is a general idea that the true communicator doesn't have notes or outlines, as he has the innate ability to communicate without any type of aid.

I have already given a large number of presentations to audiences ranging from a few people to over a thousand, and in all of them I had a plan, even when I was asked to contribute only seconds in advance.

The sense of confidence transmitted by the communication comes from practice and from the existence of a plan that guides us during the speech, as the audience can immediately tell when we do not have a plan, seeing that our speech becomes immediately incoherent.

An important note: whatever the plan may be, it must be based on the audience's previous knowledge and must have value to that same audience.

Plan format

The plan must be made the moment we clearly become aware of the reason for our presentation and we know to whom we are speaking.

It can take multiple forms:
- Topics organized on a sheet of paper.
- Detailed script.
- Mind map.

The form your plan assumes will depend on you; it is important that you feel comfortable with the format chosen.

Personally, I take the following approach:

- When time is short, because I am called upon with a few moments' advance notice before the speech, I make a small mind map.
- When it is a longer presentation, I outline a mind map and even end up making a set of notes as a guideline, but not in great detail.

Even in the context of training, I am not a follower of highly-detailed scripts, even though I understand their necessity when it comes to programs that award certificates to various groups by various trainers.

The way I see it, following a script in an inflexible manner can only be done by professional actors, but even these are remembered for moments of genius when they went beyond the script.

The story of Harrison Ford in the first Indiana Jones movie is famous, where he is confronted by an excellent swordsman and Indiana Jones off-handedly shoots him point-blank and walks away. In reality, this scene was never written; Harrison Ford was tired at the end of the day of filming and decided to act in that way.

Having a plan is great, but it must be a presentation aid and not the presentation, since we all have our moments of genius.

Mind map

The mind map is my favorite way to prepare any presentation, not only for the content, but also for the organization of the event.

The mind map or spidergram is the name given to a type of diagram, systematized by the Englishman Tony Buzan, designed for the management of information, knowledge and intellectual capital; for comprehension and problem-solving; for memorization and learning; for the creation of manuals, books and speeches, and as a brainstorming tool; and as an aid in the strategic management of an organization.

A mind map can be made on a sheet of paper with the use of a simple pen, but I prefer, whenever possible, to do it on a computer, as it allows me to quickly change the location of each item. There are Apps that allow users to make mind maps on smartphones and tablets.

To begin with, I recommend FreeMind, which has the great advantage of being free; however if you would like to go further, there are others which require payment of a fee, which allow you to do them online and export them to other applications, specifically PowerPoint.

To me, the main advantage of this method is the ability to have a complete vision of a more or less complex theme on one sheet of paper.

Image 2.01. Example of a mind map of a presentation structure using FreeMind (www.freemind.sourceforge.net)

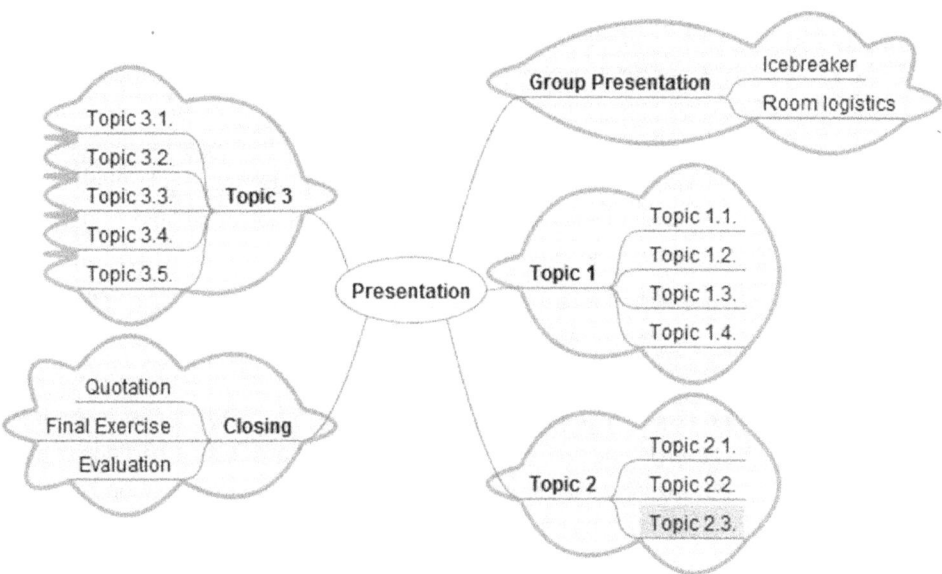

Formats

Framework

The format of our speech is many times decided by the inviting organization, in light of the objectives established for the event.

In this case, we will consider six categories of format-types:
- Training.
- Sales.
- Project proposal.
- Project report.
- Briefing.
- Speech.

Training

Training is a set of activities that have as their goal the acquisition of knowledge, skills, attitudes and behaviors required for the completion of functions appropriate for a profession or group of professions, in any branch of economic activity.

It is therefore a methodology that differs from teaching, in the sense that it specializes in professional experience, focusing on the acquisition of professional skills.

Basically, any training seeks to create a positive change in its participants and in the trainer himself, who should take advantage of each training session, seeking to gather the experience in the room and the learning stemming from the training, becoming even better.

My experience is that in giving training on a certain topic I am forced to adopt a much stricter discipline of updating and searching for new knowledge.

Trainer as catalyst

But, attention! As trainers we must have the humility to understand that we are not holders of the absolute truth, especially in an age when the volume of available information is overwhelming and extremely accessible.

The trainer or the professor is no longer the «Guardian of Knowledge», and I therefore believe it is incorrect to consider them as facilitators.

Facilitation is only possible when the information is a scarce resource with restricted access; currently, both the trainer and the professor should be seen as catalysts of knowledge.

In this sense, all the effort should be directed towards creating dynamics in the room or at a distance that accelerate the learning process, as this process is also no longer confined to the training room. Today, people are available to learn at any moment, whether by reading a newspaper article, listening to a radio program or even being in a training room.

The structure of a training action will depend largely on the pedagogical objectives to be achieved, and in structuring the training session we should be aware that in the room there will be four learning styles present:

- **Activists** – they seek concrete experience; they want to feel involved in the process; they like working on a team, as long as they are leading; and they are ready to try anything.
- **Reflectors** – they reflect over everything they observe; they like to watch others in action, compile ideas, analyzing them from various perspectives; they do not like to lead the process and are cautious.
- **Theorists** – creators of concepts, they adapt observations to logical theories; they are analytical, tend to be perfectionists, feel uncomfortable with subjectivity and think about problem solving step by step.
- **Pragmatists** – they are experimentalists, always willing to experiment with new concepts; when they learn something, they seek to test it immediately; they need real life examples; they face all problems as a challenge and are impatient.

It is worth noting that these four learning styles are present in all of us, with each of us having one that is more pronounced.

For these reasons, to keep the participants engaged in a training action, we must create content and activities that are directed towards each of these learning styles; only then can we guarantee that we will achieve the objectives defined for the session.

_{Learning styles}

Sales

The purpose of a commercial presentation is to lead the potential buyer to purchase a product, service or concept.

It should be noted that I do not encourage «imposing» a product on a buyer, using sales tactics that are more or less aggressive.

A sales pitch should be the corollary of a process of researching the clients' specific needs, the presentation being prepared from the perspective that the sale is a process of emotional transfer.

If we question anybody regarding the reasons that led them to make a purchase after a sales presentation, we will probably see responses similar to the following:
- «I trust the supplier.»
- «They seem to be people who know about the matter.»
- «I believe in the supplier's capabilities.»
- «All the doubts I had were dissipated.»

To be able to transfer positive feelings to the clients, we must fulfill an essential requirement – believe in what we are selling. For this reason, it's necessary to first convince the client that's within us, before we enter any room.

Structure of sales presentation

A sales presentation should be structured in light of prior research we have made regarding the clients' needs and even about competing offers, with their strong and weak points.

We must also have very precise knowledge of what we sell, to avoid unpleasant surprises.

Upon completing these premises, we should use a structure that follows the AIDA (**A**ttention, **I**nterest, **D**esire, **A**ction) model for commercial messages, using a structure such as the following:
- Capture the attention.
- Demonstrate the need for the product, service or concept.
- Describe how the potential buyer's needs are met.
- Explain the benefits of making a purchase.
- Include identifiable testimonials from satisfied clients.
- Present the proposal.
- Close the sale.

Project proposal

A project proposal should provide the essential information for making a decision, with the following structure:
- Underlying reason for the completion of the project.
- Explanation of the actual situation.
- Description of the facts.
- Proposal for a solution.
- Request for project approval.

In this type of comunication, it is my experience that presenters tend to start slowly, allotting their time unequally, leaving the request for project approval as implicit.

Such an approach causes confusion among the meeting's participants, who will not know whether a certain project was approved or not, since many times silence is considered as tacit approval.

Note that the level of commitment is higher when approval is requested and explicitly given.

Project report

The objective of a report is to give updated information on the development of a project.

The structure of this type of communication should be as follows:
- Brief project description.
- Reason for report.
- Those involved in the project.
- Those benefitting from the project.
- How the next phases will be completed.
- Meeting the deadlines.
- The budgetary situation.
- Alternative scenarios.

Briefing

A briefing is a short informative presentation directed to an audience who generally knows the subject.

The briefing has several applications:
- The status of a project.
- Presentation of academic work.
- Presentation of needs to a supplier.
- Presentation to clients of new functions of a product.
- Presentation of facts to the media.

Depending on the audience and the circumstances, a briefing can include the distribution of physical or digital materials. Digital materials can be delivered on thumb-drives or CDs, but they may also be made available through a hyperlink to a reserved access or open webpage.

Briefing structure

As a short presentation, the briefing can be held in a way that is programmed in time and space, but it can also be unplanned. Regardless of how it takes place, the briefing should follow the following structure:

- Purpose of briefing.
- Contents of briefing.
- Conclusions.
- Answers to audience questions.
- Close of briefing, reinforcing the points directly related to the purpose of the briefing.

Speech

A speech is a methodical exposition of a certain subject that aims to influence the feelings of the recipients.

In writing a speech, the following four points should be considered:

- Topic – can be chosen because it is of interest to the speaker or in light of audience interest.
- Audience – the nature of the audience (age, knowledge, etc.) and its position regarding the subject should be taken into consideration.
- Speaker's personality and values – the speech should reflect the person who gives it, or it will give the appearance of a sham.
- Occasion – whether it is a formal or informal moment.

Depending on the circumstances, the speech may need to be previously validated by superior entities in the hierarchy.

It can also be distributed among the media and other speakers, to facilitate everyone's job.

Structure of a speech

A speech has a simple structure:

- Orderly recognition of important people present (from the most to the least important) and of the audience in general.
- Introduction.
- Body of speech.
- Conclusions.

Two important aspects to consider in the structure of any speech:

- Transitions – the transitions between each of the speech parts will give the audience a feeling of continuity and coherence.
- Crescendo – a speech is a communication in crescendo; when the time comes for the conclusions, the audience is open to accepting them, as if they were the verbalization of their own ideas.

«And now for something completely different»

A speech does not have to be a boring experience, where a group of people sit in front of another person charged with the task of presenting something, knowing from the start that the audience will not receive it enthusiastically.

By making proper use of available technology, it is possible to captivate audiences with unforgettable presentations. This is why TED appeared.

TED is not a person; it is an event, possibly with the best speeches in the world. TED is a small non-profit organization dedicated to spreading ideas that deserve to be broadcast.

It all began with a conference that brought together people from three distinct worlds: **T**echnology, **E**ntertainment and **D**esign (TED).

Currently, the TED conferences are available for free on the www.ted.com site, where they can be seen in three ways:

- Online on the site.
- Downloading to the computer.
- Using the iTunes HD channel, which you can subscribe to and then receive new presentations periodically, which can be watched on a computer, iPod, iPad or iPhone.

One of TED's great advantages is that no speech is more than 20 minutes long, which means there is always time to expand your knowledge.

The success was such that TED has multiplied, always following the same format:

- TED Woman – focused on the role of women in the world.
- TED Global – with speakers from various countries, seeking to have a global perspective on the world.
- TED Active – meetings in a more interactive format, where the participants are able to interact with the guest speakers.
- TEDex – events organized locally in various parts of the world.

Figure 2.02. TED conferences (www.ted.com)

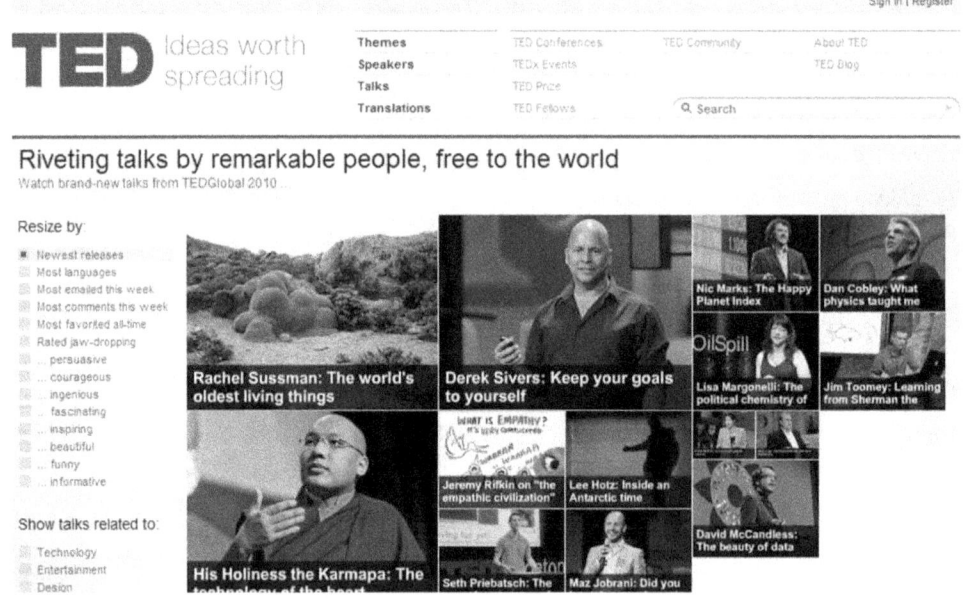

TED is unstoppable — TED created an unstoppable movement that fuses entertainment with learning and certainly will have an impact on the level of demands by audiences in the near future.

Other similar movements have risen, each with their own particular focus:

- PopTech (www.poptech.org) – presentations of 20 minutes maximum that explore the impact of new technologies on society.
- Ignite (http://ignite.oreilly.com) – 5-minute presentations on various themes, with 20 slides, each lasting 15 seconds.

Debate

Debate is the foundation of the innovation and progress of Humanity. Societies who do not foster debate are societies in decline. In some countries, the debate of ideas is an integral part of the education system.

Who I am today in professional terms is due largely to the constant debates promoted in my Ethics classes in the 7th and 8th grades of the official Spanish educational system.

Debate may be used as a methodology to create group dynamics in training initiatives or presentations, but it may also be used as a way to quickly gather the arguments in favor of or in opposition to a certain option.

In the case of organizing a debate, it is necessary to keep in mind that it should have a strict structure in terms of timing and interventions, in order to guarantee that all parties have equal opportunities to present arguments. Regarding structure, there are several schools of thought; the most representative hail from North America and Britain.

When we speak in a debate, it is essential to be aware that the final consumer of the arguments is the public watching and not our opponent who, even with our extraordinary oratory skills, will never be convinced.

Debating well...

Debating well means passing our ideas to the public, independently of our adversary's arguments. To be successful in these difficult circumstances, we should possess certain essential abilities, such as:
- Compiling and organizing ideas.
- Defining priorities.
- Evaluating proofs.
- Finding logical connections.

- Thinking and speaking succinctly.
- Speaking in a convincing manner, with clarity and impact.
- Adapting to new ideas.

These are skills that are capable of being developed. Participate in debates, promote debates and you will see that each time you will be better at debating and not just at debating, for it makes you a better speaker in any circumstance.

Just as an athlete in any sport should exercise, the speaker should debate to continually improve his performance.

Audience Analysis

Framework

Communicating effectively depends largely on the knowledge we have of the audience and how we tune that communication to the needs, experiences and expectations of that audience.

Therefore, it's important to include a more or less detailed analysis of the audience when preparing for any communication.

How to analyze the audience beforehand

In an ideal world, the organization can clearly characterize each of the participants at an event, but more commonly that will not happen; so we should resort to a few tricks to obtain relevant information, always cross-checking the sources:

- Speak with the members of the organization.
- Attend prior events directed at the same target audience.
- Watch videos of previous editions.
- Watch the impact on social networks, through commentaries, images and videos.
- Read the event site, paying close attention to all the information given.
- Determine the sponsors' target market.

- Seek, through our network of contacts, people who have attended events of the same organization.
- Gather information about previous events.

The audience analysis should always be made in advance; not doing so could certainly put the success of our communication at risk.

However, there may be circumstances that keep us from creating a prior analysis of the participants; in these cases we can:

And when there is no prior analysis of audience

- Speak with a few participants, moments before our speech, seeking to establish a profile of the audience and its expectations.
- Closely observe the audience, paying attention to clothing, composition and non-verbal communication (boredom, interest, impatience....).
- Finally, we can include small questions in our presentation that would help us understand who we have in front of us. These questions can be of two types:
  — Open-ended: of the type «What experience do you have with product X?» These have the advantage of collecting more information, but they take longer, which may not be an option in a short presentation.
  — Closed: of the type «Whoever has had a good experience with product X, raise your hand» and possibly the statement «Whoever has had a bad experience with product X, raise your hand» and then the request «Whoever does not have a formed opinion about product X, raise your hand.» For all of these options, the presenter should also raise his hand, so as to not influence the results. This will also make what is requested more visible to the audience. This method has the advantage of quickly providing some basic information from the audience, which can be used in small audiences. These questions may also be useful in large audiences, but in this last case, individual counting should be avoided, using instead the «eyeball scanner» to determine how to divide the audience in terms of percentages. If it is necessary to obtain more precise results, electronic voting equipment can be used, which we will talk about later.

Aspects to consider in the analysis	Whether the analysis is made before or during the actual presentation, we should consider the following aspects: • Demographics. • Knowledge. • Composition. • Expectations. • Origin.
Demographics	In any audience we can analyze the following socio-demographic aspects, with some being more relevant than others in each type of presentation: • Age: what age groups are represented in the room? • Gender: is the audience mostly female or male? • Family: is the majority married, single? Do they have children? • Culture: what are their social origins? • Profession: what professions are represented? • Income: to which social class do they belong?
Knowledge	In terms of knowledge of the topics covered in the presentation, we can classify the participants as: • Unknowledgeable – they know nothing or have only a rudimentary knowledge of the subject. • Curious – they have some knowledge, but it needs reinforcing. • Specialists – they have a vast knowledge of the subject and are seeking new facts.
Composition	At times, it is not enough to establish a profile-type of the audience based on demographic characteristics or their knowledge of the subject, possibly needing to go further and know who is sitting in front of us and the professional relationships and hierarchies present. This will necessitate more thorough work that might include data provided by the organization and the analysis of the following elements, as well:

- Participant files.
- Business organizational chart.
- Information on social networks.

Expectations

To understand the reasons that led a certain participant to invest the time and even money to attend any presentation, it is necessary to make an inventory of expectations, which can be done at two different times:

- Before the event – the gathering of expectations can be made using brief interviews at the time of registration, in meetings with the entities who send participants, or even by adding a field on the form, where we include an open-ended question inquiring about the participant's expectations.
- At the beginning of the presentation – if there is time, we can ask the participants directly through open-ended questions, if it is a small group. If the group is large, we will have to resort to closed questions, using systems of voting by the raising of hands or by electronic voting.

Note that the reason for making an analysis of the audience's expectations is to be able to adapt the content in light of the expressed expectations. When this analysis is made at the beginning of a presentation, its incorporation will be more difficult, unless we come prepared with several different scenarios.

Therefore, it is counterproductive to do a survey of the expectations at the beginning of a presentation if we do not incorporate it in any way during the presentation.

Origin

Added to all these aspects, it may be important to verify the geographic origin of the participants.
- Do they all come from the same country as the presenter?
- Do they come from different countries?

Dealing with international audiences forces us to take additional precautions, which will be covered in the following section.

International Audiences

Framework

As I write this section, I am returning to Lisbon from Prague, where I had the opportunity to speak at a seminar directed to partners of a North American multinational corporation. The participants came to this beautiful city in Central Europe from 16 countries located in the Americas, Africa, Europe and Asia.

In the last 5 years, I have given an increasing number of training programs and presentations in around 50 countries across 4 continents.

Many times, my audiences are not composed of local people only, but also of people who have travelled specifically to attend the events at which I am speaking, and I am certain that I have already had several thousand participants from more than 150 countries.

This international work of mine was recognized in 2008 by the JCI – Junior Chamber International, who gave me two awards: «Most Outstanding Trainer in Europe» and «Most Outstanding Trainer in the World».

I want to share this accumulated experience with the reader because I believe that, in the globalised world we live in, giving presentations to international audiences – whether in person or from a distance – will more and more become the rule rather than the exception.

Preparation

Facing international audiences requires the presenter to have a deep respect for other cultures, languages, customs and history. Prejudices and stereotypes must be abolished, and the best way to do that is to gather more information and interact with people from other cultures.

I would say that, more than anything else, if we want to be successful on the international circuit, we must assume our role as citizens of the world. Only then will we be prepared to accept the cultural differences.

Characterizing an international audience

Besides the aspects mentioned earlier in regard to analyzing audiences, in the particular case of international audiences we need to consider which of these three types of audiences we are dealing with:

- Homogenous – all the participants are from the same country.
- Heterogeneous – the participants come from different countries, or even continents.
- Heterogeneous with predominance – the participants come from different countries, but there are one or more countries or regions that clearly predominate.

Homogenous audience

The audiences in which all the participants come from the same country require an effort of greater preparation in terms of making a more thorough study of the history, geography and customs of the country, as this is a way of demonstrating respect for the participants and encouraging them to participate more actively in the success of any presentation.

Recently, I was giving training in Syria and Lebanon, which forced me to make some advance preparation on specific cultural aspects of the Middle East, but what made the biggest impact was learning during my short stay how to sign my name in Arabic, as well as how to use the Hindu-Arabic numeral system.

When working with a homogenous audience, investing in our general cultural knowledge opens many doors and hearts.

Heterogeneous audience

In completely heterogeneous audiences, where the participants have very diverse origins and where many times we only know their origin once in the room, when for example dealing with parallel sessions at international conferences, our behavior should not be much different from that in our own country, with a few variations.

Several months ago, I was contracted by a multinational corporation from the pharmaceutical sector to hold a seminar in Networking for a group of 25 talented youth in the organization, coming from all over the world—clearly a heterogeneous audience. I was careful to greet in their languages, recognizing the diversity and, at the same time, the specificity of each one.

As I gain more international experience, it is normal that during a presentation, when somebody from the audience presents an issue or asks a question, I try to find out the participant's name and country of origin, to which I normally respond, mentioning whether or not I have been to that country. This creates a type of «activators» of the audience, that is, participants who are more available, acting therefore as examples to the others.

It is important to note that a heterogeneous audience is more tolerant, as it does not expect the speaker to have a deep understanding of its cultural specificities.

In conclusion, although we are facing a diverse audience, we should seek to emphasize as much as possible the individual specificities, and the bigger the audience, the more those specificities will no longer be individual ones but of the group.

Heterogeneous audience with predominance

A heterogeneous audience, with one or more predominant countries, regions or cultures, is the most challenging, as it ends up actually being several homogenous audiences gathered in the same room.

This imposes on the presenter the same level of demand already seen in homogenous audiences, except this time it is doubled or tripled, depending on the number of predominant groups.

Linguistic barriers

I have the advantage of giving my presentations in Portuguese, English and Spanish, which are three widely-used international languages, but even so, I have already had to use simultaneous or consecutive interpreters in Russian, Polish, Romanian, Lithuanian and Mongolian.

This is one of the facts that we must take into consideration, for the world is still not flat in terms of language.

And, even when I spoke in any of the three languages that I master, I always tried to use simple worlds, free of idiomatic expressions. The objective is not to demonstrate mastery in the use of the language, but to transfer knowledge, even if only through gestures.

I think the greatest danger of not being understood is when we speak in our mother tongue to audiences from other countries or to participants who do not have the same level of mastery of our language, so that we need to speak as if we're speaking a foreign language. We should therefore:

- Use simple words.
- Not use idiomatic expressions and, if we do out of distraction or necessity, repeat the idea, but without using such expressions.
- Not make reference to situations that could only be understood by us or by those who live in our country. For example, referring to personalities from television series.
- Enunciate words well, eliminating accents.

During a speech, it may be necessary to resort to an interpreter; in which case, we should adapt our talk, taking that situation into consideration.

Interpretation

There are two types of interpreting:

- Simultaneous interpretation – interpreters are in interpretation booths and simultaneously translate what we say into the various languages of the audience, who receive the translation through headphones that allow tuning into different channels, depending on the language chosen. The interpreters also translate the audience's participation into the speaker's language, who also receives it through headphones.
- Consecutive interpretation – the interpreter is near the speaker, translating what he says and translating the audience's remarks. This method has two great disadvantages:
  — It more than doubles the length of any speech, which requires we take extra care, if not we risk ending in the middle of the speech.
  — It hinders normal communication and interaction with the audience, who will primarily see the non-verbal communication and then have access to the verbal component.

When we resort to interpretation, whether simultaneous or consecutive, we must be conscious that it will be the voices and words of the interpreters that reach our audiences.

Precautions with interpretation

The first time I was in Poland I used consecutive interpreting, and I immediately tried to create a good relationship with my interpreter, certainly a very nice and professional person.

At a certain point, we sat at the speakers' table and, as soon as we sat down the interpreter turned to me and said: «It's a pleasure to have you here»; for a split second I turned to the interpreter and started to give a response when I realized that at the other end of the table the first speaker of the day was talking. I fortunately did not say anything, but I felt like a real fool.

So, the first precaution is to be aware of the fact that you will be interpreted, but here are a few additional details:

- Creating a good relationship with the interpreters is extremely important because, unconsciously, the interpreter will put more joy into his work if he feels we respect his work and do not see him as a machine. In consecutive interpreting, this is even more important since it is not possible to translate everything, leaving room for certain interpretations by the translator.
- We should provide in advance the interpreters with materials that we will be using during the presentation, allowing them to «get into» the spirit of the presentation and have the opportunity to clarify any questions about content, since interpreters only have to be specialists in translation and not in the subjects that they translate.
- Meet the speaker – we should also provide curricular information on the speaker to the interpreters, so that they can become acquainted with him and be prepared for any references to his experience that he may make.
- In consecutive interpreting, we should use short phrases, pausing to allow for the most faithful translation of what we are saying. This presentation style must be previously agreed upon with the interpreter, as this type of interpreting requires that the interpreter and speaker be in complete alignment.

Cultural aspects

Several years ago I was invited to be the head trainer at a Leadership Academy in Russia. During the first meeting with the Academy organization, I started by seeking their opinion on several aspects, as this is the way I work.

The meeting was held by resorting to an interpreter who would translate from English to Russian and vice-versa. Everything seemed to be going well, if I listened only to the interpreter, who as any good interpreter, only translates and does not convey feelings.

Fortunately, I had gotten used to consecutive interpreting in Lithuania and Poland, and I started to observe the non-verbal language associated with the responses, only listening to the translation afterwards. I have to say that at the beginning it is very difficult, as our brains are used to processing non-verbal and verbal communication simultaneously.

But, back to the meeting, my previous experience allowed me to note that what was being said by the interpreter, although corresponding to what was being said by my interlocutors, was not aligned with his non-verbal communication, giving clear indications of concern, confusion and even fear.

I thought it best to interrupt the meeting for 15 minutes to have time to reflect, as this was the first time I had been in Russia. During this reflection, I came to the conclusion that throughout Russia's history, strong leadership corresponded to the expectations of those being led.

I returned to the meeting, this time assuming an authoritarian style of leadership, and everything started to run smoothly.

Tuning

During my childhood, when I would spend my holidays at a Spanish village near the Portuguese border, one of my skills was to tune the radio in such a way that I could clearly hear the Portuguese station. Although I was a child, I knew that the slightest touch of the radio dial could mean the difference between hearing clearly or with lots of noise.

Communication between humans follows the same principle; when we are not culturally tuned, there can be communication but the noise created is such that the message can become distorted to the point of creating serious conflict.

After my adventure in Russia, I had other experiences in countries with cultures different from mine, such as Mongolia, India and Syria, among others, and I successfully applied a certain practice.

Before arriving in each of the countries, I seek to analyze the habits, traditions and history of each, in order to «tune» my communication approach, as there is not one communication form that is valid all over the world.

Feedback

In giving training or giving speeches in about 50 countries, I have had very different types of feedback, according to the location. Let's look at a few extreme examples of positive feedback:

- In a Lithuanian school, at the end of a training session the students decided to stand and toast me with an ovation of several minutes. The professors came up to me and said, «It was good, the people liked it.»
- At a similar initiative in Estonia, the project leader came up to me and said: «There are no reasons to complain.»
- At a conference in India, after a course, several participants came and thanked me in tears, kissing my hands.
- In Portugal, at the end of a training initiative of medium duration, one participant confided to the group that she had stopped taking anti-depressants, thanks to the training.

Although at first glance they seem to be completely different reactions, they are curiously equivalent if we take into consideration the cultural differences in each of the countries.

The Baltic countries have a strong Nordic influence, where the collective is more important than the individual; therefore the recognition is, from the point of view of a Latin, very cold. If I were not in «Nordic mode», I might worry a lot about the feedback.

India is the land of gurus, who are the bearers of knowledge, and there is a beautiful tradition of respect for those who share their knowledge with others. Therefore, when somebody comes from afar to share, the respect earned is even greater.

In Portugal, we are, as good Latinos, used to freely expressing our emotions, so when we have a feedback of «it was good, the people liked it», there is cause to make us think that something went terribly wrong.

These examples illustrate well the need to culturally contextualize the feedback that we receive in each country since, if we adapt our communication to the context, we should also adapt our interpretation of the feedback we receive.

Gestures

One day I was with my children at a friend's house in Istanbul when I did a very normal game of «stealing the nose» with my oldest son, saying in Portuguese, «I have your nose!» while showing the top part of my thumb through my fingers.

My Turkish friend asked me, horrified, what I was telling my son and I naturally explained, without understanding the reason for such horror. It was then that she explained that I was giving an obscene gesture to my son, according to her culture. I could only imagine the dialogue she imagined between my son and me.

Curiously, this innocent gesture is not well regarded in Russia or Mongolia, as I could later prove. Things from the Mongolian Empire.

The gestural communication is much older than the oral, but at the same time not as uniform across the world, as we can see that certain gestures have a very different meaning, depending on the different cultures where they are made.

Sign curiosities

During a conference in Cairo, I was watching a speaker's panel where a good Brazilian friend was included and who was making a tragic error before an audience of 250 participants from Arabic countries. He had a leg crossed and showing the sole of his shoe and, even worse, moving it as if he were pointing to specific members of the audience.

Discreetly, I warned him, from my seat in the audience, to lower his foot as quickly as possible, since in the Arabic world, showing the sole of the shoe is the equivalent to «Go to h...», which certainly was not my colleague's intention.

I had the habit of doing a small cone with my fingers on one hand and hitting on the other open hand, making a noise similar to popping a cork and accompanying such a gesture with an affirmation of: «Let's go!» or «Bravo» until I discovered during a training initiative that in the countries with Russian influence this is an obscene gesture, giving a naughty interpretation to my previous comments.

In India, I was very confused because as I spoke with people I interacted with, they would slowly shake their heads as if the wind was moving them. Two good Indian friends were kind enough to explain to me that this was an almost involuntary gesture that meant they were in tune with what I was saying.

In Japan, never stab your food with chopsticks because this symbolizes death, which is not a theme that relates well to a meal.

In Muslim countries it is common for me to kiss other men on the cheek and greet women with a handshake. However if the women in question has a veil, we should only greet her if she takes the initiative to extend her hand.

In the same way that it is normal to seek to speak the language of our peers when we go abroad, when we enter into contact with other cultures we should also seek to inform ourselves regarding habits and local customs, thus avoiding unnecessary blunders, simultaneously demonstrating what any human being likes: respect.

Some additional precautions

Finally, it is very important to be aware that when we speak to international audiences, the words have a different force, depending on our origin and knowledge.

Several years ago, on a flight between Athens and Barcelona, a few minutes before landing the pilot spoke in Spanish to the passengers and said, «Ladies and Gentlemen, in a few minutes we will land at the Barcelona airport. I ask that you turn off any electronic equipment because I'm going crazy. Thank you.» In repeating the message, this time in English (with a pronounced Spanish accent), he omitted the part referencing his temporary insanity.

Up to the moment we landed in Barcelona, I thought about the complete lack of reaction by the passengers regarding a pilot who admitted, short moments before landing a plane that he was going crazy and how impossible that would be in other places on the planet.

The translation of an inoffensive expression in Spanish would be motive for chaos and despair. Fortunately, the pilot didn't translate it, giving a clear example of what can be gained and lost in a translation.

Structure of the Presentation

The success of a presentation, even improvised, depends much on knowing that the best way to communicate a message is to have a clear and simple structure.

The objective is to create, as we go through the presentation, a climate favorable to the acceptance of our ideas.

Framework

Any presentation we give should have the following structure:
- Receiving participants.
- Opening.
- Content.
- Closing.
- Follow-up.

Structure-type

Normally, receiving the participants is not considered an integral part of the presentation, and it isn't per se.

However, I include it, as I have experienced that during this phase very interesting bonds are forged with elements of the audience and data about the participants is gathered, which can be extremely useful to us.

Whenever possible we can, and should, incorporate excerpts from our conversations with the participants into our presentation. This creates proximity between the speaker and the participants, transforming some of them into our «defenders» either explicitly or implicitly.

Curiously, in an improvised speech, this phase is also important. Let's imagine the following scenario: you go to a conference when suddenly one of the speakers asks you, given your specific knowledge, to give a short speech. Two things could happen:

- If you arrived at the conference and sat down without speaking to anyone, as you face the audience you are before a group of strangers you know nothing about. You will feel, therefore, less secure, and that will affect your performance.

Receiving participants

- If you were concerned enough to arrive before the appointed time and had had the opportunity to interact with a few of the participants and speakers, you would have a feeling of playing on your home field, feeling more secure and confident.

Opening	In the opening phase we will inform the audience of what we are going to speak about, beginning this phase when we are called upon to give our presentation.

However, do not commit the error of beginning to speak even before starting your presentation, for example as you prepare the equipment (computer, projector, etc.). Be calm and begin when you are completely ready.

At this point you could use the power of silence to capture the audience's full attention. Stop, take a deep breath, observe the audience and wait to get their full attention. The interesting thing is that the seconds that you take doing this will seem like minutes or even hours the first few times you do it, but you will see that, with time, they will be effective seconds and you will have a better reaction on the part of the audience.

You can begin your presentation with:

- Something that gets the audience's attention – it could be an object, a photograph, etc., or something that has impact or is connected to your presentation.
- A relevant statistical piece of information.
- Facts unknown to the audience.
- An extreme scenario – a possible situation that is either very good or very bad, depending on your presentation and your objective. |
| Breaking the ice | As you begin your presentation, it may be necessary to create a situation in which you break the ice between the participants, as one or several of the following situations may present:

- The participants do not fully know the presenter, and vice-versa.
- The participants do not know each other. |

- The audience is tired, due to the time of day or because of a busy social schedule the day before.
- The previous presenter was boring.

Ice breakers were created for these situations, which can take, among others, the following formats:

- A small collective physical exercise.
- The participants and the speaker introduce themselves out loud or, for example, describe themselves with drawings and/or gestures.
- Make a memorable entrance onto the scene, either for its magnificence or by the absolute lack of it.
- A funny video, a song or sounds (in this case you must be careful to not hurt feelings or break the law in terms of royalties).

The choice of the perfect ice breaker will depend on:

- Imagination, but you should be careful to infuse it with something that is appropriate to the audience to which it is directed.
- The size of the audience, as it is important in regard to the feasibility of the ice breaker in terms of execution and time available.
- The type of room; visiting the room beforehand can give you ideas for new ice breakers, but it may also confirm the impracticality of doing others.
- The time available for the presentation; in courses or long presentations, an ice breaker could be an activity that takes a few minutes, but in a 5-minute presentation it does not make sense to have a 3-minute ice breaker, no matter how great it is.
- Whether it is a presentation in person or online; in the case of the online medium, I think an ice breaker is even more important due to the impersonal aspect that many people associate with it.

The content has to be simple and appealing, with the use of simple phrases.

Whenever possible, include examples, illustrations and visual aids that reinforce your message.

Content

You could use the following elements:
- Facts.
- Statistics.
- Testimonials.
- Examples.
- Comparisons.
- Definitions.
- Quotes.

You should take into consideration that people cannot stay focused for a long time during a presentation (at most between 3 to 5 minutes), so that it is advisable that you make variations during the presentation, with the hopes of holding the audience captive. For this you can use:

- Inflections in the tone of your voice.
- Questions for the audience, which can be answered out loud, by a show of hands, or even by participants standing.
- Audiovisual aids that you can intersperse during your speech.

Closing

If the first impression decidedly influences the audience's reaction, the closing will determine the dominant feeling after the presentation.

Therefore, it should:

- Summarize the presentation and clarify the most important or most complex points.
- Create an interest in the topic.
- Appeal to the participants' action, according to the presentation's objectives (do not forget that any presentation will aim to create a change).

Follow-up

As in receiving the participants, the follow-up is not normally considered a part of the presentation; however, I consider that the success of any presentation will be enhanced by a good execution of this phase.

There are various follow-up initiatives that we should consider:
- Collecting participant business cards and sending messages with:
  — thanks for attending.
  — additional details.
  — a digital version of the presentation.
- Establishing connections through social networks like Facebook or Linkedin, with the possibility of even requesting testimonials regarding the quality of the presentation.
- Sending articles, photographs or videos of the presentation.
- Sharing the presentation slides.

At some events, many speakers habitually commit a common error: immediately after their presentation, they leave in a hurry and do not stay until the next intermission, thereby missing a golden opportunity to make an almost immediate follow-up.

- We can say that there are two major objectives in this phase:
- Create new opportunities to present our ideas and, consequently, enhance our professional activity.
- Continually improve our presentations in light of the participants' feedback.

Verbal Communication

Framework

In the last centuries, a lot of importance has been given to the word, through the content of what is said and written, and there are historical moments described as turning points, derived from written documents that created religions, countries and concepts.

However, did the changes happen because of what was written? Or because of the way that it was communicated to the recipients? Can we imagine great movements in Humanity without the charisma and ability to communicate of their protagonists?

In my opinion, the written and spoken word serves as synthesis and simplification of a series of feelings and emotions that are intended to be transferred to others. But the word itself is not enough; it can only change the world when aligned with sincere emotions.

According to a study by the University of Berkeley in California, words represent only 7% of communication, the remainder being attributed to non-verbal communication, which we will cover in the following section.

«All brain» approach

When we communicate to any audience, we must be aware that each person has a different way of analyzing reality. To simplify, we can say there are two types of people:

- Emotional – they are more sensitive to emotions transferred by the words used.
- Pragmatic – they feel more comfortable when presented with facts and statistics.

It is important to note that there isn't anyone who is purely emotional or purely pragmatic; what happens is that there is a stronger bent to one side or the other.

The communicator will have a tendency to be more emotional or more pragmatic, so he will be tend to shape his communication in terms of his way of seeing reality.

This could constitute a problem. Let's imagine a communicator with a very pragmatic approach, speaking to an audience with a great emotional inclination; we will certainly have a situation of poor communication.

Therefore, it is necessary that those who communicate be conscious of how they analyze reality, adapting it to the way their recipients see the world; only then will it be possible to create empathy and consequently facilitate communication.

Speaking to the senses

Neuro-linguistic Programming (or simply NLP) is based on the idea that the mind, body and language interact to create each individual's perception of the world, which can be altered by the application of a variety of techniques in which verbal communication has great importance.

According to NLP, there are three ways that people capture and analyze information:
- Visually – sense of sight dominates.
- Aurally – sense of hearing dominates.
- Kinesthetically – touch, taste and smell dominate.

Each person has one type that is more developed than the others, so we can speak of three types of people that can be identified by the types of expressions and words they use in their communication.

Type of people	Expressions	Words
Visual	«I look favorably on...» «I don't see how...» «You're a sight for sore eyes»	See, look, show, perspective, image, clear, clarify, light, dark, brilliant, colorful, visualize, illuminate, vague, imprecise, distinct, foggy, a scene, horizon, bright, photographic.
Auditory	«This is music to my ears...» «The dogs bark, and the caravan moves on...» «This does not sound good...»	Hear, speak, say, listen, ask, dialogue, agree, disagree, sound, noise, rhythm, melodious, harmonious, symphony, cacophony, yell, roar.
Kinesthetic	«This doesn't smell right...» «This is a situation that really touches me...» «I feel comfortable in this situation...»	Feel, touch, connection, relaxation, concrete, pressure, sensitive, insensitive, tender, delicate, solid, firm, immobilized, soft, hurt, connected, warm, cold, tension, hard, excitement, weight, relief.

As communicators, we tend to communicate more with our predominant senses, but we create empathy with an audience when we communicate to all types of people, therefore we should practice the use of expressions and words that appeal to all those who make up our audience and not just to those who identify with our way of seeing reality.

Telling a story

One of the most effective techniques for transferring complex ideas is to tell a story. There are several reasons for this:

- From the time we were very young, we have been accustomed to stories as a way of basic learning, connected with a concrete reality.
- Something that is common to all cultures is the existence of storytellers, characters who use this structure to transmit new ideas and concepts, who helped the formation of religions, political ideals, nations, etc.
- Stories, due to their sequential nature, aid the memorization process.
- When we are told a story, we immediately start to visualize situations, protagonists and their emotions.

Precautions with stories

In communicating with multicultural audiences, I have become aware of the importance of telling stories to aid in the transmission of information and experiences. However, a few precautions must be taken:

- Do not invent or tell other people's stories, unless you are a good storyteller.
- Use your own stories, which can come from ordinary day-to-day situations. You do not need to be an adventurer of the seven seas to have good stories.
- The story should connect to what you intend to communicate; never tell a story just because you want to tell it.
- The story should also have a connection to the audience; therefore use terminology that is accessible to all.
- If it is our own story, we tend to do the long version since we are personally talking about our favorite topic. We should, therefore, be brief, without getting lost in the details.
- Be sequential; don't jump around backwards and forwards along your story line. If necessary, make a small pause in the narrative, to introduce any contextual information.
- Involve the participants in the story, questioning them with: «How many of you have felt the same...», «I'm sure you know this type of person...», «You can imagine what happened...».

- Prepare a conclusion to the story that is connected to the message that you intend to include in your communication.

It is important to note that it is not possible to tell a story well without using non-verbal communication to facilitate the task of the recipients' imagining the situation.

The power of the word

Although we have seen that the word in and of itself represents little in the communication as a whole, this does not mean that it cannot be powerful, many times having an effect contrary to what we were aiming for.

The spoken word can be shaped by our non-verbal communication, so that it is possible to use assertive language that, modeled with an adequate non-verbal expression, will not create conflicts.

The problem with written words is that they are very difficult to model, because their interpretation will not only depend on the words used but also on the reader's mood.

It is not by chance that many conflicts are created or escalated because of the interpretation of what was written.

You can do a simple experiment, using a recorder:
- Read aloud a message from somebody that you like.
- Pause.
- Read aloud a message from somebody that you do not like.
- Listen to what you recorded.

You will be surprised at the differences in intonations, the reading speed, the pauses and emphasis on details and the errors. As emotional beings, we introduce our own emotions; therefore, we should be careful when it comes to communicating in writing.

Beginning

For years I have been explaining to people that are close to me that they should never approach me with phrases like: «I have bad news...» or «I have good news...»

In the first case, after the first words, my mind starts to immediately rush to anticipate the news, and I enter a process of suffering, imagining each of the possibilities, and at the end of the communication may come to the conclusion that the news was not that bad at all, feeling somewhere between relieved and furious for the unnecessary suffering.

In the second case, expectations are created that may or may not be met, with the good news actually becoming bad news.

These are extreme examples of words that we use to begin a communication, and it is very risky to initiate any communication anticipating the feelings that the audience will feel after the communication.

Good starts

Good ways to initiate communication are:
- Telling a story.
- Introducing yourself.
- Explaining why you are there.
- Presenting an unknown fact.

Apologizing

One of the most common mistakes made by speakers is the need they feel to apologize for anything and everything, as a way of avoiding their shyness or of avoiding a genuine confrontation of ideas, seeking to appease the audience, but achieving the opposite effect.

Unless there truly are reasons for a brief apology, such as a delay caused by the speaker himself, communication should be fluid, doing away with any type of crutches such as successive apologies: «I trust my dear colleague will excuse me...» This will make the audience feel that the speaker is not confident in what he says, and he will, therefore, no longer be interesting.

This is good, but...

«But» is a very dangerous word that is often used. It deserves special attention as it can completely kill the message we intend to give.

Try a simple exercise: pay somebody a very nice compliment and end it with a «but» and a pause, and ask that person what he felt. Certainly, the person you chose will feel very confused.

Using the word «but» restricts creativity. How many times have our suggestions encountered that type of barrier: «We could do this that way, but...»?

So, I prefer to systematically substitute the word «but» with the word «and».

Try to substitute «but» for «and» throughout an entire text you have written. Read it again and you will see something very different. Before, the text presented problems, now it presents solutions. As a recipient which do you prefer: to only be presented with problems or to be presented with solutions?

Non-Verbal Communication

For millions of years, Humanity lived without any system of codification of ideas, emotions and concepts, with communication taking place non-verbally.	Framework

With the advent of languages, verbal communication started to gain importance, but even so, non-verbal communication continues to represent more than 90% of communication.

And everything becomes more extraordinary if we consider that the great majority of educational systems focus on the mastery of verbal communication, paying little attention to the non-verbal component of communication.

Therefore, it is important to realize that we can improve our non-verbal communication, because any small improvement will have great repercussions in the way we communicate.

Our non-verbal communication is commanded by the subconscious; for this reason its analysis is very important in interrogations, when one intends to discover if the person being interrogated is telling the truth.	The subconscious at the wheel

On the other hand, it is very difficult to coach non-verbal communication, when it comes to people who do not follow careers of acting or spying, in which a substantial part of their training is centered on mastering non-verbal communication.

In this section, it is important to understand that sincerity is a value of extreme importance when we communicate, because the recipient analyzes our sincerity by continually aligning our verbal communication with the non-verbal.

Connection between sub-consciousnesses

The way our recipient analyzes communication is not primarily conscious, that is to say, he consciously listens to what we say, but all our non-verbal communication is analyzed by his subconscious.

The reason for such behavior is simple; our conscious mind is capable of processing little information simultaneously, while our subconscious is capable of processing ad infinitum simultaneously, reaching conclusions that will be assumed consciously, even if we do not understand the underlying reasons for such conclusions.

I bet, for example, that you have been through this situation: one day you spoke with someone who said everything good that you wanted to hear about a business, person or product, but after your conversation, you made a comment such as: «There is something not quite right about what he said.»

happened was that your subconscious captured a misalignment between the verbal and non-verbal communications, placing the sincerity of the interlocutor in question.

To better understand this form of communication, we will see what we use to communicate non-verbally.

Posture

Imagine that we are waiting our turn to resolve a delicate situation that requires the solution of a specialist, and a staff member with drooping shoulders shows up; would we be confident?

Another situation: we are waiting for the person who will lead a reconciliation meeting and suddenly somebody barges in who looks as if he has swallowed a coat hanger; would we be calm?

The examples given seek to illustrate how the body is a communication tool that we must know how to use and align with what we want to communicate.

For this reason, there is no ideal posture for all moments and audiences. Our posture should be shaped to the communication that we seek to establish.

For example, if at a certain time we desire a certain level of empathy from the audience, it could be interesting to lean our body slightly in their direction. However, if we want to underline a point with firmness, we must assume the most upright posture possible.

Also, do not speak to an audience with your arms crossed, hands in your pockets or behind your back, as this will reflect a defensive posture or even of somebody who is hiding something.

However, this does not mean that you cannot have those postures. Try the following: stand straight up in front of a mirror, with arms crossed, and analyze the feeling that transpires from that posture. Then, tilt your head slightly to one side, in order to make the corresponding part of the neck more visible.

You will see that now the feeling is different: you went from authoritarian to understanding, even with your arms crossed, because you are demonstrating openness to the recipient.

The reason for this feeling lies in the following: the neck is one of the most vulnerable parts of our body; when we make it accessible, we show that we trust the recipient and all good communication is based on mutual trust.

Gestures

Gestures are powerful, especially when we are being filmed or photographed, because they can be used as summaries of all that we communicate.

When we thumb through any publication, we find that reporters seek to align moods that were photographed with the content of the articles, that is, if a reporter writes that a speaker was perplexed about a certain situation, there will be a photograph of a perplexed speaker; if, on the other hand, a picture is shown of a confident speaker, the reader would naturally be confused.

Gestures also have an important cultural component:

- There are gestures that are normal in some societies and offensive in others.
- Societies in the south of Europe especially appreciate speakers who use very open and theatrical gestures.

- In contrast, in countries in the north of Europe or the Far East, gestures in communication should be more contained.
- In any culture, we should take a gradual approach, that is, start with smaller gestures and evolve to more open gestures, but always considering cultural differences.
- In the same country we may have very different audiences in terms of reaction. Imagine, for example, your gestural communication at a bankers' conference and at a conference of university students; I bet that you do not think you will take the same approach.

In spite of what has been said here, do not try to consciously change your gestural communication during your communication; this could call into question the sincerity of your words. The best approach is to prepare beforehand, analyzing the audiences and internalizing the most appropriate behaviors.

Our subconscious directs our gestures, so that when we try to do them consciously, we will appear phony. It is not by chance that actors have the expression «getting into the character's skin», referring to the automating of a set of communication dynamics, in order to align words with gestures.

Tone of voice

Napoleon Bonaparte, Alexander the Great, Julius Caesar…, all these historical figures were brilliant strategists in their time, having a voice capable of mobilizing great armies.

In their time, there were no effective ways of amplifying the voice, so the effort had to be greater. They had to have a tone of voice that could convince the first rows of the troops to advance without fearing certain death, thus encouraging, by their behavior, the rows that followed.

Tone of voice represents around 30% of communication. To have an idea of how powerful a communication tool it is, choose an excerpt of any historical speech and read it in a monotone and monochord voice – you will certainly put your audience to sleep.

Tone of voice can and should be practiced; I leave here a few aspects to take into consideration:

- Projection of the voice, which is the ability to place it in such a way that we can be heard by large portions of the audience without needing to yell. To achieve good voice projection, we should practice in an empty room, preferably with one person at the back, who can tell us if they hear us in good conditions.

- Voice projection is not only useful for being heard by the whole audience, it also conveys confidence to the audience.
- Inflexions in tone of voice serve to emphasize certain points in what we say. They are the equivalent of bold in written communication. It also causes the audience to always be attentive. You should be very clear what your key communication points are and use inflexions at those points.
- Tone of voice is also associated with good diction; otherwise a message runs the risk of becoming confusing. Two simple pieces of advice: avoid words that you have difficulty pronouncing and, if you must use them, practice, practice, practice.

The power of silence

(...)

You are surely wondering what the three dots are above. It is a form of demonstrating in writing the power of silence.

Many presenters, when called to give their presentation, try to not lose time, starting to speak right away, even before they are ready to start their speech.

The power of silence can be used at different times in communication:

- **Before starting** – allows the speaker to «feel» the audience, get absolute silence in the room and recognize the audience's presence.
- **During a presentation** – when we want to emphasize a certain point or briefly reflect on a certain issue. In this last case, many speakers start to speak and try to reflect on the issue at the same time, which does not work well; others, afraid of the silence, use the annoying «uh! uh! uh! uh!».
- **To conclude** – when we want to end a presentation, a small moment of silence immediately before helps the audience to focus on what is important.

I should warn you that, when we work in silence, there is a difference of perception of time between the speaker and his audience, for example in an initial 3-second silence:

- Speaker – depending on the degree of comfort, each second will seem like a minute, creating a certain discomfort to the speaker, if he is not well trained.
- Audience – there will be very different perceptions, as some participants will be talking to each other and will only be aware of silence in a room when it is completely quiet, so it will naturally be shorter than 3 seconds, with some thinking that the silence happened suddenly.

Being conscious of these different perceptions will allow the speaker to use silences more effectively, without any discomfort.

Eye contact

In communication, eye contact serves to affirm our convictions, but also to maintain a connection with those we are speaking with.

As parents, how many times have we told our children to look at us when they talk to us? This is the way to educate them towards effective communication, since we do not believe in the sincerity of someone who speaks and does not look us in the eye.

In this area, shy people have great difficulties: in fact, one way that these people face audiences is by looking everywhere but at the people, and with this behavior they call into question what they intend to communicate.

As a recovering shy person who now deals daily with audiences of various sizes without any nervousness, I can say that it is possible to overcome shyness. How? I will give a simple piece of advice: If we think of not doing something because of shyness, we should do it and not think any more about it and so on and so forth.

But how to establish eye contact with all the participants in an expanded audience? Rationing our gaze? No, we should use the lighthouse approach.

yourself as a lighthouse, that your eyes are the lights, and that the audience is the area that you aim to illuminate; what you will do is sweep your gaze over all the audience, in a regular, but non-mechanical way, thus causing each participant to think that the speaker is looking at him.

Controlling non-verbal communication

As we have seen, language is and should be controlled by our subconscious, or else we will appear phony.

As it is not possible to open our head and have a cup of tea with our subconscious, how can we improve our non-verbal communication?

Using video can improve our communication; we should record ourselves on video and later analyze the way we communicate. In this way we can see what we do best and worst as we communicate.

Video is a powerful tool in helping us improve our communication methods (verbal and non-verbal), and as such, there have to be certain precautions in carrying it out:

- The recording should be as long as possible. Besides having more material to analyze, it will allow the elimination of certain behaviors associated with being conscious of being recorded.
- Do not speak to the camera – only highly trained people speak to the camera as if it were a person.
- Try to capture the sound with a microphone so that you can hear your voice, with all the inflections.
- As you record, use wide angles. This way you will be able to see the speaker's full body and even part of the audience.
- Watching the video should not be done on a small screen. The ideal would be to use a projector and screen so we can have a larger-than-life image and consequently more access to the details.

In spite of video being an extremely useful tool in the self-analysis of how we communicate, we can and should gather the maximum contributions from our audiences, whether through face-to-face communication or through the analysis of tweets made by participants during and after our presentation.

The Room

Framework

The room is a very important element, but the choice of the set up may be ours or imposed.

If it's ours, we should choose the format based on the objectives of our communication.

If it's imposed, we must adapt, conscious of the limitations, overcoming them with imagination because, as we well know, nothing is definitive.

I cannot help but tell a small story at this point. Several years ago, on a Sunday, I was sleeping and I dreamt that I received an emergency call to substitute for a trainer who had experienced a sudden problem. I went to give the training in an amphitheater, and things were going so poorly that I had to apologize to the trainees, saying that I was sick.

At the start of the next morning, already awake, I received a phone call from the manager of a training company for which I worked several times. Surprisingly, I was asked to urgently substitute for a trainer who had had an accident, with the manager offering to take me in his own car to Oporto.

During the trip, I joyfully described in detail my nightmare, not imagining what was awaiting me. When we arrived in Oporto, we went straight to the training room, which in no way corresponded to the one in my nightmare but almost immediately, the local person responsible for logistics informed us that we would need to change rooms and go to the amphitheater.

I looked at the manager to whom I had described my nightmare and catching a glimpse of terror on his face, I responded with a smile, «Good thing I'm not superstitious!»

Contrary to my nightmare, the training went very well, with only one curious detail: on the return trip, we had to stop at almost all the rest stops along the motorway, as I was feeling very ill.

Room analysis

It is advisable to do a prior analysis of the room where our presentation will occur. This analysis does not need to be long, but it is convenient to keep in mind the following points:

- **Set-up Style** – we will see later the implications of each set-up style.
- **Acoustics** – we should check to see if there are echoes, if any sounds from other rooms or from the exterior are audible, and what kind of voice projection we should use.
- **Lighting** – we must see what natural and artificial lighting there is in the room and specifically on the screen, if one is being used. We should also check to see how the lights are controlled, since many times the lighting of the room is made through fairly complex commands.
- **Room entrances** – ideally not on the side where the presenter will be.
- **Planned space for the speaker** – will there be a podium, a table or merely an area of circulation for the speaker?
- **Chair layout** – will there be lateral and central aisles between the rows of chairs to facilitate participants entering and exiting?
- **Water** – should it be a long presentation, it is important to have water. Some practical advice: prefer bottles that can be closed after drinking, since glasses of water near computers are not a good combination.
- **Support staff** – we should aim to speak with the people who will give logistical support to our presentation, for example, registering participants, helping with the sound and other equipment. These people are rarely recognized for their vital work in the creation of a unique experience in the room. I have personally experienced that merely a brief exchange is needed to have them as fabulous allies, capable of going above and beyond what is expected.
- **Room equipment** – we must confirm that the previously requested equipment is present, such as:
    — Projector.
    — Screen.
    — Flipchart or white board.
    — Sound equipment, should the number of participants make it necessary.

Set-up styles

In this section, we will cover the more common set-up styles, namely:
- Theater.
- Auditorium.
- Banquet.
- Classroom.
- Meeting.
- U-shape.
- Standing.

Theater

In a theater-style room, the speaker is the center of attention, as he is standing in front of the audience, which is seated in chairs that may or may not be attached to the floor.

The speaker normally has a stage allowing him to be seen by all the participants. If there are many attending, small TV screens along the wall or screens on which the image of the speaker is projected can be used.

Should you feel comfortable and it is possible (there is no accentuated unevenness and the sound allows for it), the speaker can leave the stage and make his speech closer to the participants, even walking among them.

Walking among the participants can be something fun for the speaker and for the audience but the following precautions need to be taken:

- **Sound** – everyone needs to hear what is being said, therefore hand or lapel microphones are essential at these times.

- **Distraction** – you should do it in moderation; if you walk too much, the audience will stop concentrating on what you intend to communicate and start trying to guess where you will be going next.

- **Eye contact** – when you walk through the audience while you are speaking, you naturally lose eye contact. It is not mandatory to keep eye contact 100% of the time, but you should avoid turning your back to the audience. It is preferable to advance and retreat, but always maintaining eye contact with as much of the audience as possible.

FACE-TO-FACE COMMUNICATION

- **Direction of walking** – you should avoid walking in one direction only, as you might create in the subconscious of a part of an audience that they are less important than the part where you walk.
- **Filming** – if we are being filmed for image projection in the room or for later watching, we should avoid sudden changes of direction or running, because the person filming does not know what we will do next. We can walk, but we should do it slowly, and before changing direction we should stop and begin walking slower. Otherwise, we will make our audience in the room or at home seasick.

Image 2.03. Diagram of room set theater-style

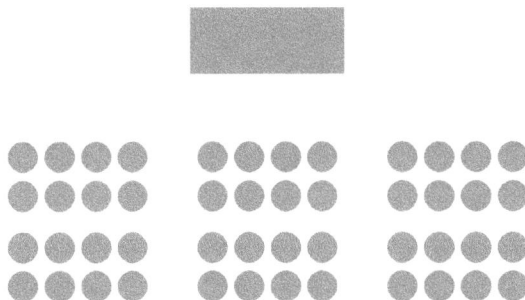

Auditorium

The layout of an auditorium is similar to the theater layout, where the speaker is the center of communication, but it presents a few variations:

- **Chairs** – are fixed, not allowing for the creation of group dynamics.
- **Speaker's position** – due to the difference in levels found in this type of room, the speaker still appears in greater prominence, but with much more restricted freedom of movements.
- **Stage** – usually, the space for the speaker is less flexible and smaller in area, favoring speeches with little interaction with the audience.

Image 2.04. Diagram of room in auditorium format

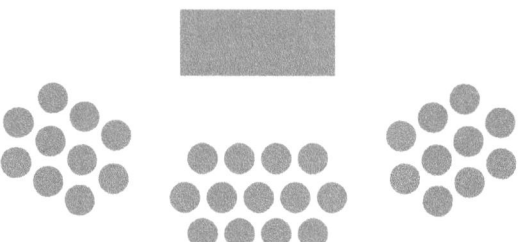

Banquet

As the name indicates, the banquet-style layout functions as it does at banquets, where people are grouped at tables, either in assigned seats by the organizers or with open seating.

In either case, this set up facilitates the creation of group dynamics, not being suitable if the use of such activities is not planned, since it will create an expectation that will not be met.

Some precautions to take with this set-up style:

- **Line of sight** – all participants should have the same opportunity to establish eye contact with the speaker, without having to move their chairs, so the places with the backs to the speaker should not have any chairs. The exception to this rule is for banquets in which all the seats have to be filled, for logistical reasons, with the speeches then needing to be shorter. In the case of large banquets, screens should be used around the room so that all guests may have the same opportunity to see the speakers.

- **Tables** – the most recommended type is round, but square tables or two rectangular tables could be used to create a type of island. We need to keep in mind one detail: tables should not be parallel, but with their center pointing to where the speaker is planned to be, allowing people to sit in front and slightly to the side.

- **Materials** – if the participants need various materials (pens, paper, models, etc), these should be distributed beforehand to each group, to facilitate work and time management.

Image 2.05. Diagram of room set banquet-style

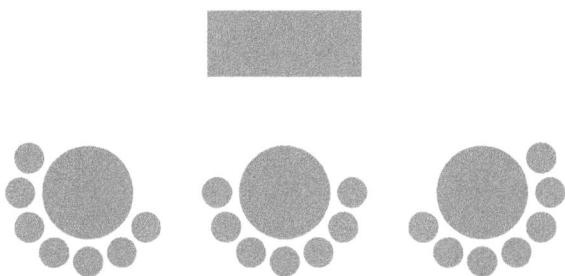

Classroom

Classroom style is the one most commonly used in schools and universities; the inherent logic is that the professor is the holder of knowledge and the students are mere recipients.

Questions of room design, the number of participants and inertia mean that this continues to be the most typical model in education, but I believe that such a format will gradually be abandoned due to the following limitations:

- It does not foster group work.
- The professor is at the center of the process.
- It perpetuates the habits of taking notes during the class, very appropriate if the objective is memorization of concepts, but perfectly inadequate if the objective is to understand the concept of the presentation.

However, this classroom set-up style can be very interesting in training initiatives where it is necessary for the trainees, in small groups, to use parts, or equipment, reproducing what the trainer is demonstrating or creating new parts, equipment, units, etc.

Image 2.06. Diagram of room set classroom-style

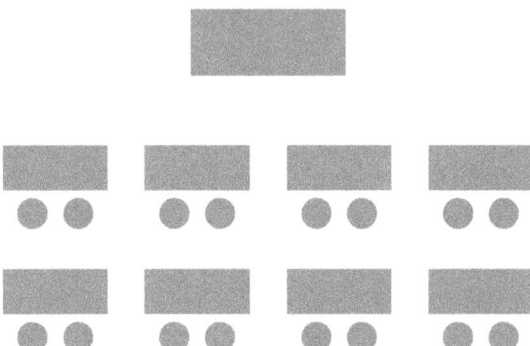

| Meeting room | The typical meeting room set-up, that is, the long table with people seated on both sides, is not suitable for more than 15 people. |

This set-up is ideal for discussing ideas that are presented by all or by designated speakers, with the room functioning as a cohesive group in terms of work dynamics.

There is one variant that is seen at summits, where it is possible to triple the number of participants, even providing each participant with the option of having a small support team. In this variation, the tables are set in an oval, circular, or square shape, leaving an open space in the middle, where decorative elements and screens for watching the presentations can be placed.

But for this to function correctly, it is necessary to have:

- A sound system allowing everyone to be heard by all the participants.
- A moderator with power to cut speakers short.
- Screens that can be placed on the tables or, better yet, on the floor in the middle open area, thus allowing, at all times, a visual contact with each speaker in turn.

Image 2.07. Diagram of room set meeting-style

The room set in u-shape has become the typical room for training in contrast to the classroom style.

U-shape

This custom stems from the advantages that this set-up brings:

- All participants are on equal footing, motivating them all to participate.
- All can maintain eye contact.
- A specific area for the speaker at the top of the U that simultaneously allows him to move to the interior, getting close to all the participants.
- The majority of rooms at training centers and hotels are prepared to accommodate this set-up style.

However, we must not think that the U-shape works for everything, as it does have a few limitations that we should consider:

- Dividing the group into work teams is sometimes complicated because it involves moving tables or walking with chairs in the air.
- Work groups tend to be the same, for reasons of proximity in the room among the participants.
- The process is highly centered on the speaker.
- It is not appropriate for groups larger than 20 people.

Image 2.08. Diagram of room set in U-shape

Standing

In situations where time is scarce and there is a need for quick conclusions, we may need to make a presentation in which all the participants remain standing.

In this type of presentation, it is important to remember that after a certain amount of time, participants will be uncomfortable, with two factors adding to their discomfort:

- The time that the speaker takes.
- The level of participation: the more participation between participants, the less they'll notice the passing of time.

For these reasons, we should take the following precautions:

- Position ourselves in such a way that everyone in attendance can see us; if necessary use something to raise us – stage, box, chair, etc.
- Not walk from one side to the other, to prevent the audience from doing the same and losing visual contact.
- Be concise and direct.
- Seek to create some participation from the audience, preferably with closed questions.

This set-up style could serve as a complement to all the styles previously described if it is intended to create work groups.

Image 2.09. Diagram of room where participants are standing

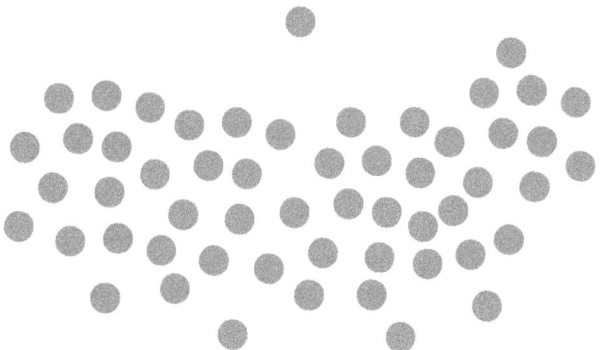

Chairs

In a room, the type of chairs may determine the type of presentation to give in light of the advantages and limitations they may represent. Therefore, we'll talk of three types of chairs:

- Movable – in this case we have complete flexibility in creating group dynamics and of even making new room set-ups.
- Fixed – they may be fixed to the floor or chairs attached in chain, creating aisles. In both cases, the use of group dynamics will be affected.
- With support – this type of chair facilitates note-taking and even completing certain necessary tasks during a presentation.

Support rooms

When there are large groups, there is a type of support room, normally designated as break-out rooms, which are rooms where each work team can gather to receive additional instructions and work on the preparation of a document to present in the main room.

Resources

Framework

Nowadays, any presenter has a line-up of resources at his disposal that can be used; many of them are currently computer-based, which allows for creating stimulating presentations.

On the other hand, it is easy to find speakers who are fascinated with a new resource, seeking to use it, regardless of the circumstances surrounding their presentations.

Choosing resources

So, anytime we think of using a resource, we should confirm whether its use is appropriate, in light of the following variables:

- Communication objectives.
- Audience profile.
- Room set-up.
- Level of experience and knowledge of the resource we want to use.

Audio resources

Audio resources are related to sounds in the background, from people, or from music that we can use.

Music, for example, can create a favorable environment for the upcoming communication. The level of sound should be raised, but not as if in a discotheque.

To prepare audiences, I customarily use dance music (for the younger ones) or lounge music (for the less young). With programs like Windows Media Player or iTunes, it is possible to create lists of specific songs (playlists) for each type of event, or within the same event, for a particular moment, even using music at the end of the presentation.

Today, it is very simple to bring sound to a room even if there is already sound equipment in the room. MP3 players, iPods, smartphones, tablets and computers can facilitate the use of sound in any presentation.

The problem may be in the sound reaching everyone, even when it is a small audience. In the cases of larger audiences or poor acoustics, it will be necessary to use sound equipment or have people responsible for running the sound we are using. In this last case, it is necessary to agree on signals beforehand, to communicate the start of sound.

In the case of smaller audiences, I used to ask the organization for speakers for my computer; recently however, I have discovered the wonder of mini-speakers, which have four great advantages over regular ones:

• Size – they fit in the palm of your hand.

• Totally portable – they do not need electricity or batteries.

• They have a small internal battery that is rechargeable through the USB cable, that is, you can use it and charge it on the computer itself.

• The sound potential is extraordinary for its size – I have used them for groups of more than 30 people, without any problems.

Image 2.10. Example of an X-mini mini-speaker (www.x-mini.com)

Visual resources

Increasingly, visual resources are multimedia resources, with the laptop being a true multimedia center that allows us to combine video and audio in an easily accessible way.

With the advent of YouTube and similar sites, it has become very easy to give concrete examples during a presentation. But, be careful. Although we have increasingly better access to the Internet, it is best to always have a plan B up our sleeve!

There are browsers like Firefox or Chrome that have added features capable of downloading YouTube videos onto the computer or to external drives.

Overhead projector

The overhead projector is a piece of equipment that had its golden era before the use of PowerPoint, but we cannot say that it is dead, actually quite the contrary.

It is obvious that we must not use the projector when we can substitute the expensive and inflexible transparencies for slides in a presentation program.

However, we can use it anytime we would like for work groups in a session to present their conclusions; we need merely supply transparencies and appropriate markers to each group and we will have faster and more objective conclusions than if we had asked for PowerPoint slides. I know, from experience, that regardless of the type of people that are in the groups, there will be a tendency to spend 80% of the time on format and 20% on content of the slides, with transparencies having the inverse proportions.

But projectors have modernized themselves as well. Currently, we have projectors that include cameras, allowing the presenter to handle objects of various sizes, with their images projected onto the screen so that a large audience can see the movement.

Image 2.11. Example of a projector with incorporated camera (www.topex.com)

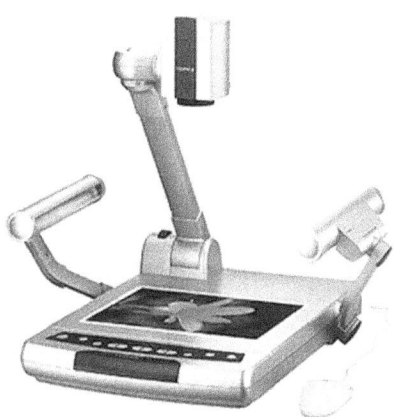

Flipchart

Simply, the flipchart is a support for a giant pad that allows us to write and draw for various audiences.

The great advantage of a flipchart is its flexibility; at any moment, the presenter, reacting or not to a stimulus from the audience, can take the opportunity to complement the presentation and write a few words or graphics.

The flipchart can also be used as a central resource in a presentation, even substituting the slides if, for example, no projection is available.

Flipchart sheets can be prepared beforehand or can be made during a presentation, using contributions from the actual audience.

As an alternative to flipchart pages, it is possible to use a whiteboard that has as a disadvantage not being able to remove created sheets for later use during the presentation.

Image 2.12. The Flipchart

Advice for using a flipchart

Some practical advice when using a flipchart:

- Type of handwriting – it should be of a size and style that allows easy reading from any point in the room.
- Horizontal and vertical layout – the content should be displayed uniformly on the sheet.
- Writing – if we are creating the flipchart content during the presentation, we should stand back as we write and orally repeat what we write.
- Audience involvement – the flipchart is a great resource for involving the audience, there is no better way to do that than to gather their input and present it before all, for example with a flipchart.

- Key words – we should not write complex text; we should write topics with key-words.
- Facilitate – give the audience time to read and occasionally copy.
- Markers – before starting, it is necessary that all markers to be used be in good condition. Many times I get to rooms where I find an acceptable quantity of markers, but few write decently.

We can add 3 small tricks, that can be of great use:

- To avoid memorizing all that we need to write, you could write a few key words in pencil in the corner of each flipchart page, as a memory aid.
- It is also not easy to draw quickly, so it would be preferable to prepare the drawing in pencil and at the appropriate time, trace over the pencil revealing the drawing. In training in Syria, I surprised the participants when I wrote, in front of all, «Welcome» in Arabic (my assistant had written it in pencil beforehand).
- We can be a little more sophisticated and create pretty flipcharts, using a very simple technique:
    — Place the flipchart in front of the projector.
    — Project the image that we would like to draw onto the flipchart.
    — Draw the images with the markers, tracing the projected lines.

PowerPoint

PowerPoint has become the presenter's best friend, but we all know slides that have killed presentations.

PowerPoint is a program of utmost usefulness, but it should be used intelligently, bearing in mind that a presentation is not a set of created slides, because if it were, the presenter's presence would be completely dispensable.

Any presentation in PowerPoint or similar programs should serve as an aid and anchor to the work done by the presenter, avoiding at all costs «Death by PowerPoint».

When it comes to similar programs, there are several alternatives to Microsoft PowerPoint on the market, such as:
- Open Office Impress.
- Apple Keynote.
- Google Docs Presentation.
- Prezi.

Advice for using PowerPoint

Some practical advice when using PowerPoint:

- Get PowerPoint training and explore its various functionalities. During a recent research I found around 50,000 tutorials on YouTube regarding this program.

- PowerPoint has a variety of absolutely extraordinary animations. My «favorite» is the one called electrocardiogram. But don't dazzle. Use simple animations that show the content at the moment you need it, in other words, using what we used to call «Mask» on transparencies.

- If the animations are varied, the font types are even more so. We should also avoid the overwhelming majority of fonts, opting always for the most visible:
    — Verdana.
    — Tahoma.
    — Arial.
    — Times new roman.

- One of this program's great advantages is the ability to use images that allow us to communicate emotions, but two precautions need to be taken:
    — The images used should be relevant to what will be presented; otherwise we will find ourselves with a distraction.
    — The images might have associated royalties, so it is advisable to use: our own images, free image databases, PowerPoint clipart. If necessary, purchase images.

- As was stated for the flipchart, key words and topics should be used. Complete phrases will be pertinent in the case of quotes.

- KISS (Keep It Simple, Stupid) – one subject per slide and without a lot of text. This implies dividing complex content over several slides and avoiding unnecessary embellishments.

- No more than 4 different colors per slide, or else we will have a rainbow, which will only distract.

- Do not use abbreviations, unless they are widely used by the entire audience.

- Whenever possible, use the pointer from the mouse and avoid the laser pointer as nobody can maintain the dot completely still, unlike the cursor, and it distracts.

- Use the «Presenter view» function. This will allow the audience to see the slides normally, while the presenter simultaneously sees:
    — The slide that is being presented.
    — The sequence of slides before and after.
    — The duration of the presentation.
    — The current time.

- Use the keyboard:
    — If you press the B key, the screen will turn completely black, as if the projector has been turned off, for example, to do a small exercise. Pressing the B key again will return you to the presentation.
    — If you press the W key, the screen will turn completely white.
    — If you input a number, 118 for example, followed by «Enter», you will immediately go to slide 118.
    — In the slideshow mode, if you press the F1 key, you will have access to these and other shortcuts.

Non-linear presentations

Imagine that you must give a presentation to an audience, of which you are completely unaware of what they know or do not know regarding the information you will present. They could be specialists or they could be unaware of the most basic concept that you will use.

In this case, you can use action and hyperlink buttons on slides or other files, which will only be shown depending on the audience needs.

That is, instead of making a linear sequence of slides which appear one after another, you will instead have, for example, on slide 4, a connection to slide 1 or a file that is not in the sequence.

Prezi

It may be that PowerPoint dominates the presentation market, but for that very reason audiences appreciate it even more when presenters bring something different.

Prezi is a breath of fresh air in this field, allowing one to create animated presentations with great impact due to a zoom animation of the slides.

I consider this program to be very useful, especially in presentations where it is necessary to present complex concepts in a brief manner.

Prezi has paid versions for the computer and a free version where it is possible to create presentations online and then download the final result in a flash file to your computer.

Image 2.13. Prezi – a different way of giving presentations (www.prezi.com)

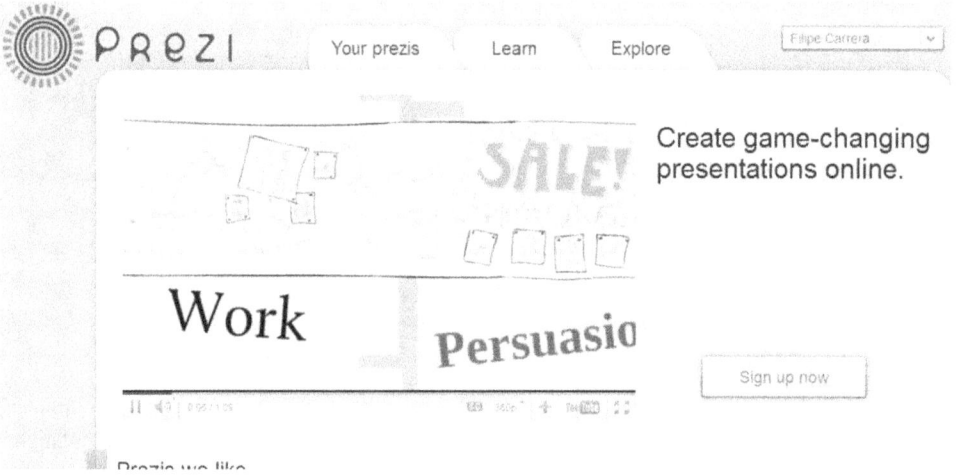

Mind map

Our brain works by associations and comprehends complex concepts better if it can visualize them in a single image.

Mindmapping technique is highly utilized in brainstorming sessions, project planning, etc. It is not used as much in presentations, due to the monopoly of PowerPoint.

My experience is that a simple mind map can be more useful when we want an audience to clearly understand the implications of an idea, the steps to develop a project, etc.

A mind map can also be used in gathering input from an audience.

Image 2.14. Example of a presentation using a mind map (www.mindmap.com)

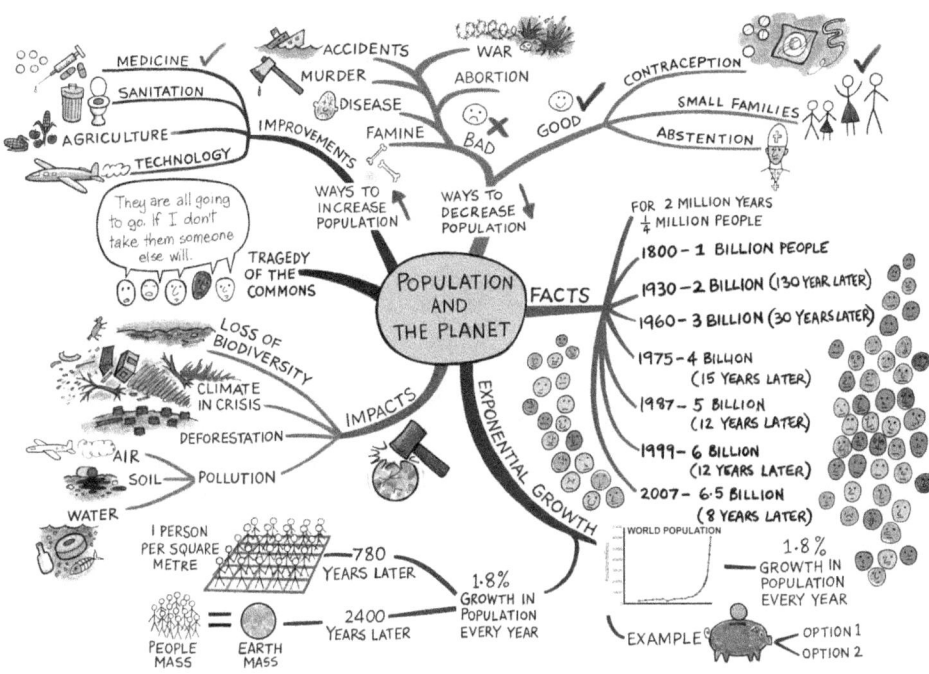

Support Equipment

In this section, we will cover some equipment available that will allow us as presenters to completely master a presentation without the need for people to assist us, which many times is not viable.

There are many types of equipment, but I chose those that in my opinion are essential or that allow a more natural approach.

Framework

The graphics tablet, used by designers, has become available to anyone and, for less than 100 dollars it is possible to have a good graphics tablet that can serve a very useful purpose for any presenter, if you do not have a computer with a touch screen.

Graphics tablet

One question that is frequently asked when seeking input from the audience is what to do with that information. We have already talked about the creation of transparencies or flipchart pages, but we want to go further and control the process better, using fewer resources.

The graphics table can be a brilliant alternative, following the steps below:

Step	Description
1	Install and connect the graphics tablet to the computer where the presentation will be made, using PowerPoint, for example.
2	Reserve blank slides for the moments when your plan calls for gathering contributions from the audience.
3	Place in slide show mode.
4	When you get to the blank slides, activate the stylus.
5	Using the stylus from the graphics tablet, write the audience's input which, thanks to the projector, will be visible to all.
6	Should you need to gather conclusions from the session, it is possible to save the input at the end.

Image 2.15. Example of a graphics tablet (www.wacom.com)

The wireless presenter remote is an inseparable companion to any presentation using my computer, as it allows me to walk around the room and even surprise my audience by controlling a few multimedia elements of my presentation.

There are a few more simple ones on the market that basically only allow moving to the next or previous slides. There are others with multimedia capabilities, which function as a mouse.

I prefer the latter; by the way, in the image below you can see the wireless presenter I use, which has the following functions:

- The transmission of data is made via radio, which allows me to move up to 15 meters away from my computer without losing control of my presentation or having to point my antenna at the computer.
- It includes a laser pointer that I rarely use as it is not very practical when we want to point at something for longer than a second since we cannot keep the pointer absolutely still.
- It includes a mouse function that allows me to use the cursor as a pointer and even click on hyperlinks to open pages or other files.
- It does the normal function of moving forward or backward in a slide presentation.
- It auto-runs, which allows me to use it on any computer. When I have colleagues making presentations with me, they usually ask to borrow my presenter.
- The antenna (also visible in the image), is stored inside the device next to the battery compartment.
- Although it holds two AAA batteries, it only uses one at a time, with one being a back-up in case the first one runs out.
- It has a volume control, which allows me to place music or any other type of sound before, during or at the end of my presentation, at the exact moment I want it and at the desired volume.
- It has the capability to block buttons – this function is very useful when we want to place the presenter in our pocket and do not want to involuntarily start the slides.
- There are also two keys that allow us to completely black out the screen and start or finish our presentation.

Wireless presenter

• Finally, the design is very well thought out because it has a series of grooves that prevent the presenter from slipping out of our hand, ideal for those days when our hands sweat a lot.

Image 2.16. Example of a wireless presenter (www.targus.com)

Video projector

The first video projectors were the so-called «data-show», a transparent plate connected to a computer and placed over an overhead projector that required the room to be completely dark to be able to see anything. In its time, it was an extraordinary innovation, since the printing of good transparencies could be the biggest cost of any presentation.

Then video projectors appeared, equipment that was very large and very costly, and few organizations had the luxury of owning one, causing the rental of these devices to flourish.

We can say that this type of equipment is now widely available, due to the drastic reduction in cost (from several thousand dollars to a few hundred), with the image quality improving, regardless of the lighting in the room, and with the reduction in size, this has become a truly portable piece of equipment.

On this last point, the big innovation is the pocket video projector, shown in the next image, allowing new possibilities. This type of projector is not appropriate for large audiences, but it can be of great use in presentations to audiences of up to 20 people, depending on the lighting in the room.

Image 2.17. Example of a pocket video projector (www.3m.com)

The video projector is an immensely useful device, but I have noticed that most of the time it is underutilized, so I will leave here a few pieces of advice for more effective use.

Advice for using video projectors

- The screen onto which it will project should be tightly stretched and visible from all angles in the room.
- The projector should be very level, centered and at a 90º angle facing the screen.
- The projector should be placed discreetly so it is not the center of attention. Many times, the projectors are fixed to the ceilings in rooms or behind the screens. In this last case, it is necessary to use a screen specifically for the purpose.
- The projector cable should be protected so that nobody trips over it, since suddenly turning it off could, at the least, reduce the lifespan of the projection lamp, and in the worst case, burn it out. It is important to note that this is a special bulb and, therefore, very expensive. This is why there is a way of turning the projector off gradually.
- Many presenters, not knowing the PowerPoint function of blacking out the screen, place paper over the projector to avoid a projection. In so doing, they block the ventilation and may cause a fire.
- We should be careful to not walk in front of the projector's light beam, lest we create a type of shadow theater. By the way, never look directly into the light, as there is a real danger of blindness.
- We can connect a video camera or a DVD player and not only a computer to a video projector. This equipment normally has connections for sound and, sometimes, the capability of reading external memory.

- In general, all projectors are accompanied by a remote, which 99% of people don't see any use for so they put it aside. I am a fan of a function that is present on all of them, under the name «freeze» or «pause». This function allows me to maintain the projection of an image that I have chosen and simultaneously perform other tasks on the computer, such as opening other programs or even trading computers. It is extremely useful to remove that less than professional image of seeing the work environment of a desktop in the brief interval between presentations.

Electronic voting equipment

In a session with a large audience, it may be interesting to integrate the input and opinions of those in attendance; however, its size could be prohibitive since, if in an audience of 100 people each person took 30 seconds to express an opinion on a particular topic, you would have more than 50 minutes invested in that one point only.

Fortunately, there are already alternatives – electronic voting equipment that allows one to ask a large audience closed questions (yes/no, multiple choice) and integrate their responses directly into the presentation.

For example, in a presentation, we can ask how many people suffer from headaches and even attribute a scale: 1-never, 2-rarely, 3-occasionally, 4-often, 5-always.

The participants have devices in their hands by which they can choose one of the five alternatives and, in mere seconds, the presenter has access to the results in absolute format, in percentages, and represented through graphs, being able to immediately share them with the audience through a PowerPoint slide created for this purpose by the software associated with the electronic voting equipment.

Image 2.18. Electronic voting equipment (www.powervote.com)

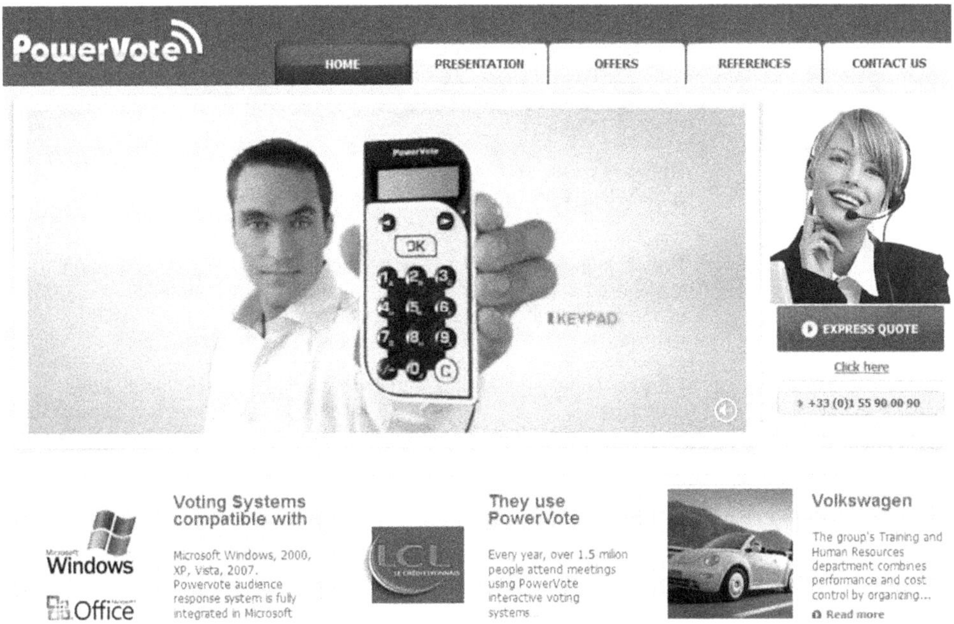

Dealing with Difficult Situations

Framework

I have given thousands of presentations in almost 50 countries, across 4 continents, in very different venues and in front of very distinct audiences.

From my experience I have learned that flexibility in action and thought are of extreme importance in these occasions, as we are confronted with totally unexpected situations that force us to change the course of a presentation in order to avoid a disaster.

I have presented here a few of these situations, with useful advice for each; for others, not covered, I have a piece of general advice: prepare well and use good common sense.

Difficult participants

Sometimes we are confronted with participants who for one reason or another like to challenge the speaker in a problematic way. Here are a few situations and possible solutions.

If...	Then...
They ask questions out of context to the presentation	Do not respond directly, so as to not lose the rest of the audience. Say you will respond during the break.
They ask questions that in reality are small speeches to demonstrate their knowledge.	If possible, use them as allies during the presentation, giving them a little recognition, but forcing their next contribution to be shorter.
They continually make side comments.	In the case of a small audience, you should directly ask if there is a question or comment to share.

Equipment malfunction

In 2010 I had the opportunity to visit the magnificent Roman theater of Aspendos (Antalya – Turkey), built during the reign of emperor Marcus Aurelius. I was very impressed with its acoustics, as my voice was perfectly heard in any of the 12000 seats available, without any use of sound equipment, which is truly remarkable.

Now, any presentation can make use of an unlimited option of equipment to make it more effective and memorable; however, we must be prepared for malfunctions and breakdowns, but always with one thought in mind: the presentation is made by the presenter and not by his equipment.

If...	Then...
The sound equipment is giving feedback.	This means we are getting too close to the speakers and our microphone is picking up the sound from the speakers. We should check the sound before the presentation, to define where we can circulate with the microphone on.
The sound equipment is not working in perfect conditions and makes noises and/or is cutting out.	Since these types of problems irritate any audience, and even more so if the speaker insists on continuing, it would better to stop using the microphone and start projecting the voice well.
The video projector does not communicate well with our computer.	Several solutions present themselves (from fastest to slowest): 1. Use a computer from the organization. 2. Restart the projector, with the computer on. 3. Reduce the computer monitor resolution and check that the option «Duplicate screen» is chosen and not «Expand screen». 4. Restart the computer with the projector cable connected.
The video projector does not work, or the light burns out during the presentation.	In this case, preparation means everything; the best option is to have an extra projector, but that may not be possible, so we should be prepared to not use the projector.
The computer does not work.	Having the files on external memory is the ideal for these times of trouble.
The Internet does not work.	As a general rule, we should avoid using the Internet during a presentation, as «Murphy's Law» is always present. Some possible alternatives: 1. In the case of YouTube videos, there are programs that allow these videos to be downloaded to the computer. 2. In the case of wanting to show online navigation of an application, we can record the steps between pages, using programs like Captivate, Camtasia Studio, Camstudio or Wink.

Problems with files

A particular case is the files that can be used. Today, the excuse of lost files does not work anymore, due to the number of available alternatives.

As I travel a lot, sometimes to places more or less dangerous, I must consider all the possibilities, and so I share here my security measures:

- Besides computer files, I always have a back-up copy in my office.
- Up to a short while ago, I used to take with me a pen-drive or memory card in my mobile phone, with the files.
- Finally, I place my files online, using the Intranet or a service like Dropbox, which allows me to access my files and, if necessary, download those moments before the presentation.

I should say that, because I am always in motion, much of my work is done in widely separated places and, before using Dropbox, I was worried about the possibility of losing files. Dropbox has removed this worry, as it has created a folder on the computer that synchronizes with an identical folder online every time that computer connects to the Internet.

Image 2.19. Dropbox: an online back-up service (www.dropbox.com)

Unexpected audiences

Before any presentation, I try to speak with the organizers to understand the profile of the expected audience, but sometimes the organizers are completely wrong.

Once, at a seminar in Lithuania, I was informed by the organization that the participants would mainly be local young people. When the time came for the presentation to start, I noticed my audience was composed of retirees.

How to act in one of these situations? It is obvious that we cannot give the presentation as planned, ignoring the change in circumstances.

We should adapt the presentation (possibly shortening or lengthening a few points) and the speech to the audience present.

So that this can be done at the last minute, it is necessary to combine three decisive factors:
- Flexibility.
- Knowledge of subject.
- Preparation of several presentation alternatives.

Almost empty room

Sometimes, the organization of an event reserves a room for 300 people. We then expect that many people come and hope that nobody will be kept from the session.

However, once it is time for the session to start, there are only 6 people in the room. This is a demoralizing situation, in the sense that the presenter tends to fixate on the 294 people who did not show up.

In a case like this, we need to concentrate on the 6 people who did attend and decided to invest their time and perhaps their money to listen to us. The first step will be to try and make a circle of chairs with the participants, where we include ourselves, and give a presentation as if it were a conversation.

I know, from experience, that this approach is motivating for the participants and for the presenter.

3

MULTIMEDIA COMMUNICATION

Collaborative Platforms

Framework

More and more, any professional must use new ways to communicate, resorting to Information and Communication Technologies; otherwise, he will quickly become obsolete and hardly productive.

The inherent costs of meetings in person are no longer admissible, as they can be substituted with alternatives that are much cheaper and more effective.

In this point, we will cover the collaborative platforms available to communicate more and better.

What are they?

There are many collaborative platforms; I list the market leaders as examples:
- Adobe Connect – www.adobe.com.
- Webex – www.webex.com.

Features

In general terms, these platforms have the following main features:
- Use of participants' audio and video.
- Chat window.
- Whiteboard sharing.
- Host work desktop sharing.
- PowerPoint presentation sharing.
- Immediate or scheduled meeting appointments.
- Meeting recording.

Image 3.01. Webex, a market leader in collaborative programs (www.webex.com)

Preparation

Many face-to-face meetings fail due to faulty preparation and failure to follow schedules.

When we hold a meeting using a collaborative platform, the factors that influence its success or failure are, with a few variations, the same as those as for meetings held in person:

- If it is the first time that some participants are using a collaborative platform, they should enter the system 15 to 30 minutes early, depending on the number of participants who are doing so for the first time.
- Should the participants be used to the interface, they should enter 5 minutes before start time.
- There should be a well-defined agenda, with meeting start and end times.
- It is necessary that the moderator have the power and authority to give and remove the opportunity for participants to speak.
- The participants should be those who are strictly necessary, and if the need arises, it should be possible to call others to participate, who must be forewarned of this possibility.

During

We must be aware that the use of these types of platforms continues to be something new and the majority of people consider communication through these methods as unusual and even frightening.

For these reasons, it is beneficial to take the following precautions during the conversation.

- The participants should be aware that there is no visible non-verbal communication that allows others to understand who would like to speak, leading to the need to establish simple rules to request to speak. On these platforms, it is customary to request to speak through an icon that appears on the other participants' screens, placed there by the person making the request.
- Whenever possible, the feature that shows the image of all the participants simultaneously should be used. In addition to personalizing the meeting, the feature discourages multitasking.
- There is an enormous temptation for participants to take advantage of dead time, no matter how small, by checking e-mail, checking pages, chatting or even answering the phone, so the meeting should be kept at a high pace, with constant request for participants' feedback through the platform's tools that allow online voting.

Afterwards

One of the great advantages of these types of platforms is the possibility of making available to those who participated, and to all who need the information, a recording of the meeting, which functions as detailed minutes of all that was presented, discussed and decided.

The moderator should be held responsible for sending the hyperlink of the meeting recording to all those interested.

Practical applications

The collaborative platforms could advantageously substitute many face-to-face communication events, such as:
- Meetings of various types.
- Mentoring sessions.
- Product presentation.
- Brainstorming sessions.

Skype

Framework

As an inveterate user, I could not resist dedicating a few pages to Skype, which, for me, is an indispensable tool in my day-to-day life, whether for personal or professional use.

What is it?

Skype, since its release in 2003, has specialized in voice communication over the Internet, with video communication also becoming available in 2006.

Currently, Skype functions are similar to those of Microsoft's MSN, but with a better acceptance within the professional market, due most likely to a more trustworthy voice and video service and also to a better developed component of calls to landlines and mobiles. In the case of mobile phones, it also allows sending SMS texts.

Another of Skype's interesting features is the capacity to make conference calls, where up to 24 people can participate simultaneously, regardless of being contacted or not by Skype, phone or mobile.

It allows for desktop sharing, letting one user see another's desktop, which makes it much easier to explain how to do something on a computer that we are not seeing directly.

has also put out a lot of effort to overcome problems in sound quality during calls.

Note that, independently of the program, a better or poorer call quality is due in great part to the sharing of resources on our computers.

For professionals who feel those problems, I strongly advise the use of headsets and microphone with an incorporated sound card, as these will be able to achieve sound quality superior to that of a normal phone call.

Skype is truly a program capable of transforming our computer, tablet or even smartphone into a communications hub, thanks to the diversity of services it offers.

Image 3.02. Skype: a communications center for businesses and professionals (www.skype.com)

Personally, I use a combination of Google docs and Skype as a collaborative platform, which allows me to:

One way to use it

- Hold teleconferences with up to 24 participants simultaneously.
- Share files in real time and even alter them as a group, using Google docs.
- Videoconference one-on-one (in the free version) or between up to 10 people in the paid version. But you can do it for free using Google hangout on Google+.
- Write instant messages, which can include hyperlinks.
- Share my desktop.
- Communicate with non-Skype users through their landlines or mobile phones.
- Send SMS texts.

Image 3.03 .The paid version of Skype permits video calls between various users (www.skype.com)

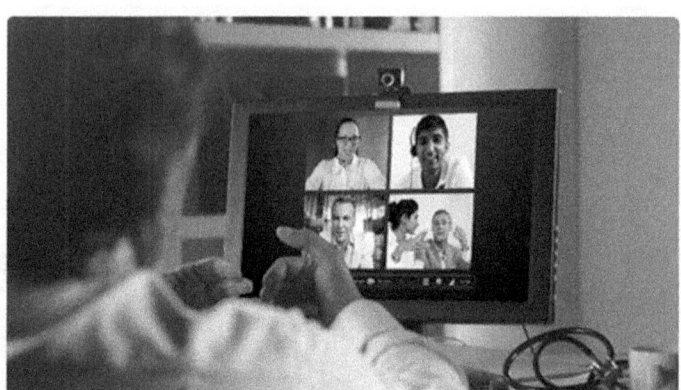

Advantages	To me, Skype is a wonderful program and I cannot understand companies who forbid its use. Its advantages are clear and I will list only the main ones:

- Reduced costs – calls made between Skype users are free and calls to mobile phones and landlines are at reduced rates.
- Ability to inform third parties when we are available or not to receive calls, avoiding unnecessary interruptions.
- Possibility of sharing files and workspaces during a conversation, saving time and tiresome explanations.
- Good video call image quality.
- Possibility of speaking with various people simultaneously in real time (for professional reasons, I held a teleconference every month between the USA, Netherlands, Portugal, Togo and Macau, and it always worked well).

Practical advice	Skype is a tool and, as any tool, its usefulness depends on the way it is used, and so I list a few pieces of practical advice:

- Invest in a good headset; it will certainly pay for itself. Sound quality is extremely important in this type of communication; and being a headset, it leaves our hands free to use the keyboard whenever necessary.

- I love laptops since they already include webcams, and progressively better ones, so the video call should be used, except when the bandwidth does not allow it. Do not worry, because Skype will warn you when the bandwidth is insufficient.
- Try to establish contact with your collaborators, suppliers, clients and partners over Skype; you will see an increase in your response speed and quality.
- With a video projector (or large monitor), a good microphone and speakers connected to the computer, you will create a video-conference room.

Are chats just idle chatter?

In my opinion, chats are not just a nuisance, they are counterproductive. In the beginning, instant messaging was extraordinary, due to the lack of audio or video alternatives.

Currently, I see organizations that start to introduce their own systems of instant messaging as if they were a great innovation, at a time that they should be introducing systems like Skype.

Using chat promotes what I call switchtasking: while I write, my communication partner has an irresistible temptation to do something and the same will happen when it is his turn to write on the chat, resulting in the following:

- We both lose more time than we would have with an audio conversation.
- We exponentially raise the possibility of poor interpretations of what we are writing, since the level of concentration during the conversation is greatly reduced.

The interesting thing is that the majority of people have the perception that they are saving time, but then do not understand why days go by and the work is magically accumulating.

Attention! I am not against using a chat. On the contrary, I think it is extremely useful in the following cases:

- To complement a voice conversation; it does not make sense to spell complex words or hyperlinks when we have chat capabilities.
- For conversations of extremely short duration and with the use of emoticons, which allow an increase in the non-verbal communication component in chat.

- To ask somebody if they are available to talk or to respond that we will only be available in 5 minutes since we are on the phone.

The mystery of the video call

In 2010, I was one of the many passengers in Europe who were stranded due to the Icelandic volcano Eyjafjallajökull (try to read that without hesitations ☺). I had booked a launch session in the city of Targu Mures for the Romanian version of my book *Networking – Professional Survival Guide*.

The organizers were desperate; there were guests, media, and everything was booked and in motion. Given the complete inability to fly, I suggested in a very natural way that I could give my presentation over Skype.

To me it was a natural suggestion, as I use video calling in Skype daily and consider it a sufficiently reasonable substitute for being there in person.

other great advantage is flexibility. On the Romanian side, they merely needed to connect the computer to an LCD monitor with an HDMI cable, place some speakers, and there I was, in spite of Eyjafjallajökull.

Videoconferencing has never been so accessible, and yet it continues to be a mystery to the majority of professionals. For this reason, the next point is dedicated to this «mysterious» form of communication.

Videoconference

Framework

I remember watching, in the 80s, a television documentary on a project that was done in a small French community. All the homes in that community were connected by a videoconference system that allowed neighbors to communicate with one another through their televisions.

I believed that a revolution in the way we communicate at a distance was very near. Almost 30 years later, I find something curious:

although easy, accessible and affordable videoconference systems exist, the overwhelming majority of people rarely use this type of service, if they have used it at all.

It seems incredible that it is so rarely used, although we have communication technology at our disposal, which 30 years ago we could only dream of while watching science fiction series such as «Space: 1999» or «Battlestar Galactica».

The scarcity of its use has to do with a set of mental blocks that greatly affect our use of this technology:

Mental blocks

- We are not used to seeing our image on the screen, this in spite of the advent of photography and digital videos.
- When we see our image on a screen, we fundamentally associate it to our personal life and very rarely to our professional life.
- The «me» we see and hear does not correspond to the «me» we think we are.
- We consider it a breach of privacy when the other person sees our image. This mental block is particularly funny: do we ever have meetings in person wearing burkas?
- Consciously, we do not give importance to non-verbal communication, preferring other types of long distance communication methods, such as telephone or chat.

I will approach videoconferencing as a communication concept regardless of the method used, of which there may be many:

Range of coverage

- Skype, MSN and other platforms that allow video calls over the Internet.
- Webcast, as a transmission of audio and video using streaming media technology. It can be done over the Internet, corporate networks or Intranet for the broadcast of this type of content.
- Mobile phones with video call capabilities, namely mobile phones of the 3rd generation and higher.
- Systems of video calling supported by smartphones and tablets.

- Television sets with video calling capability, namely through Skype.
- Traditional videoconferencing systems with directional cameras and microphones.
- Telepresence – a sophisticated videoconferencing system in which the participants gather in rooms designed for this purpose. Given the image and sound quality, they have the feeling that they are in front of one another.

As we can see, there is no shortage of alternatives, from free to extremely costly, from totally portable to permanent, from low sound and image quality to a better quality even than that in reality. There are no excuses to not use it!

Image 3.04. Example of a Cisco telepresence system (www.cisco.com)

Preparation

Besides the normal matters to consider in preparation for a meeting or presentation, the following specific precautions should be taken with a videoconference:

- The files made available during a videoconference should be in a specific meeting folder on your computer.

- A test should be conducted the day before, if it is the first time to use a specific platform, as this will allow you to not only test the platform itself, but also the actual equipment (camera, microphone,...).
- Whenever necessary, confirm that technicians are available as back-up to give assistance should something happen.
- Plan the activities and their execution – this is a videoconference, not a video speech. You could promote debates, brainstormings, group dynamic initiatives, etc.
- When a doing a videoconference for the first time, it would be good to perform some tests with free programs a few days in advance, which will allow you to internalize the dynamics of a presentation using videoconferencing.

During

At the launching of my book in Romania, which I did using Skype with video and a split screen to show some PowerPoint slides, I planned an activity for an audience of around 50 people.

I had initially planned to do this activity in person, but because of the famous Icelandic volcano, I now had the challenge of doing so 3000 km away.

It was a group dynamic exercise that started with a volunteer and then required the entire audience to move. When the volunteer offered himself, I asked him to get closer. Fearfully, the volunteer came closer to the television that was speaking to him and was telling him to get closer, and then I asked everyone in the room to start the movement I had planned.

The impact was extraordinary since the audience had the expectation of a videoconference presentation without any interaction and all of a sudden they were doing things that the man in the television was asking, watching them as they did it.

On another occasion, I was invited to give a presentation at a university in Oporto. In checking my calendar, I saw that I was supposed to be in Kiev, Ukraine, that very same day. The normal response would be: «I cannot!»

But since I am flexible, I asked: «Could it be via videoconferencing on Skype?» The response was positive and on the day and time set, I was in an office in Kiev with my computer giving my presentation to a well prepared room in Oporto.

However, I did not limit myself to merely showing my image; with the split screen function, I went through slides and at the end of the presentation I opened up a time for questions. To alleviate the participants' tension, I kept giving indications to the room assistants where the questions were being asked, which gave a clear feeling of interaction.

In conclusion, although it is a videoconference presentation, it does not have to necessarily be expositive and boring, as we can see in the examples given. With a little imagination, we can bring some interactivity that can define the success of a presentation.

Afterwards

This is many times the most important phase in the creation of true change; unfortunately, I see that many times the effort concentrates on the preparation, guaranteeing the necessary technical conditions for operating the presentation, followed merely by the relief of having been able to do it.

Here are some activities that can be done after a videoconference presentation:

- Online availability of the videoconferencing recording to participants and all interested absentees.
- Sending of a draft document with main conclusions.
- Sending of files presented and other supplemental files or hyperlinks.

Applications

As I see it, videoconferencing can advantageously substitute a part of presentations given in person.

We will see, with the passing of the years, organizations imposing more and more restrictions on travel and the inherent costs related to presentations.

There are a lot of concrete videoconferencing applications; here are a few examples:

- Training where the trainer cannot bring specific equipment to the presentation; for example, laboratory equipment or even guinea pigs.
- Recruitment of collaborators, for two sets of reasons: the collaborator to be recruited needs to be prepared to communicate via videoconference. Also, if I interview somebody on a videoconference, projecting the candidate's image on a screen, it is possible to better observe facial micro expressions, that is, I have a better ability to analyze non-verbal communication.
- Multipoint videoconferencing, allowing people who are geographically distant and have complicated calendars to gather quickly, avoiding the usual logistical problems.

Vodcast and Videos

Vodcast is a way of distributing videos over the internet, or over a network of computers, to create a list of streaming videos that is updated automatically as new videos are inserted on an Internet page.

Framework

For the purposes of this type of communication, we will consider not only videos produced for platforms such as iTunes or YouTube, which allow subscriptions in order to receive updates, but also videos placed online through other platforms that are based not on subscription but on sharing, as is the case of videos placed online through Google docs.

The common availability of wide band is altering web surfers' behaviors. No longer mere readers of content on a computer screen, more and more they are becoming investigators of multimedia content, not only on the computer but on other devices, such as portable mp3 and mp4 players.

The multimedia world

Additionally, what would be the best way to know a professional? From a set of texts or from a multimedia approach, where we see photographs, audio clips and videos?

If we are not yet using these methods, it is important to start experimenting and learning how to use them, in this way gaining new skills that will certainly be of great use in the medium term, as I believe that, just as it is important for many professionals to master presentation techniques in a room, shortly it will be very important to master multimedia presentation techniques.

Obviously, we live in a multimedia world and we have not transferred that aspect to the Internet for two reasons:

- Bandwidth (problem already overcome).
- Mental barriers, yet to overcome (but we will see how we can do that).

Mental barriers

But if there are only advantages in the use of multimedia to transfer our message, why is its use the exception and not the rule?

In my opinion, it has to do with the following mental barriers:

- A video is a lot of work to make and comes at an elevated cost – currently, it is actually very easy to create a video: software is free, digital cameras have never been so cheap and, if we so choose, even a mobile phone or a webcam can be used, so that creating a video is accessible to any person, which is easy to see through the various examples on YouTube.

- Professional actors are needed for multimedia content – this mental block has to do with the issue of comparison. It is very interesting to see professionals, who have great ease in transferring knowledge to their peers, become intimidated when a camera starts to record. The big problem here is the attempt to imitate actors or news anchors. In the use of a product, the most important thing is not to have a recognized actor explaining how to use a product, but to have someone who is not an actor be able to provide client support at any time through video explanations, which will be recorded on the product page, for future access to other clients.

- It is faster to make a text than to create multimedia content. A well done text, all-encompassing and motivating, takes immeasurably longer than a multimedia presentation, since written text calls for extra care and always thinking of the circumstances in which the reader finds himself.

These barriers will seem even more absurd when our telecommunications use primarily images, since we still see the recording of voice and image as something more appropriate to actors than to common people.

Several types of equipment may be used for video production: Equipment

- **Video camera** – there is a great diversity of formats (HD, 3D). In choosing this type of equipment, always consider how the sound will be captured, since it is an essential element of a video.
- **Webcam** – this will be useful in quickly explaining a procedure or sharing a piece of knowledge, but the image and sound quality may not be the best.
- **Smartphone or tablet** – the advantage of this type of equipments is its dissemination. At this time, we can say that it is present in any part of the world, since then capturing and sending videos has become something very simple. The disadvantage could be the sound and image quality.
- **Camera with video function** – a little less disseminated than mobile phones with video functions, but usually with a little higher quality and storage capacity.
- **Computer for editing video** – with software for making edits.
- **Tripod** – in various sizes and formats, their purpose is to get a stable image; however, there are video cameras on the market that already guarantee a very stable image, even when filming without a tripod. Small flexible tripods that allow placing the camera anywhere can also be used.
- **Microphone** – given the importance of sound in video production, we will cover this equipment in more detail in the following point.

When necessary, we can go above and beyond the microphone that comes with the equipment and use an external microphone, especially if we are recording outside or if we want to capture wide angles that allow us to observe the movements made by those being filmed.   Microphone

There are several types of microphones that we can use:

- **Directional** – at times incorporated into the cameras, but it can also be a separate accessory. This type of microphone captures the sound depending on the direction in which the camera is pointing; it is not very suitable for distances beyond 2 to 3 meters since it begins picking up background sounds, but it does allow for a certain mobility of the person being filmed.

- **Wired** – this type of microphone is connected to the camera and placed near or even on the clothing of those being interviewed. The sound quality is very good, eliminating background noise, but it limits the interviewees' mobility, which may cause complications when they need to complete tasks for the recording.

- **Wireless** – this is a microphone that allows complete mobility of those being interviewed and the filming of wide angles not in the direction of the speaker, since the camera can capture sound through the receiver, at more than a dozen meters away. There are two types of solution:

    — Adapted through traditional wireless microphones that are used in conferences and which connect directly to the camera through the sound input.

    — There are video camera manufacturers that already supply wireless microphones as accessories. In the following image is a microphone that uses Bluetooth technology, allowing high quality recordings up to 15 meters away, even when filming outdoors.

Image 3.05. Example of a Sony wireless microphone, with receiver, using Bluetooth technology (www.sony.com)

Software

Normally, webcams, cameras and video cameras come with brand specific software for video editing and converting into formats that allow editing by other programs more appropriate for video editing.

There is a great variety of video editing software, but for issues of simplicity, we will cover three programs accessible to any user:

- **Windows Live Movie Maker** (www.microsoft.com) – This program is directed towards users of Windows applications and allows for fast and intuitive editing of video content, permitting for example:
    - Recording content directly from webcam.
    - Adding title and credits.
    - Cutting parts of the video.
    - Creating automatic films.
    - Recording to multiple outputs: DVD, disc, mobile devices, YouTube.

- **iMovie** (www.apple.com) – This is a program with the same functions as the previous one, but it is more sophisticated, as is the case of these types of programs that work on the Apple operating system.

- **Camtasia Studio** (www.techsmith.com/camtasia) – This is video editing software (not free) that, besides the usual video editing program functions, allows the creation of files in various formats and easy integration into web pages.

Creation process

The video creation process will go through the following phases:
1. Script creation.
2. Location and wardrobe selection.
3. Video recording.
4. Video editing.
5. Final result.

Script creation	To make the process faster and easier, the script should not follow the cinematography model, as we are not dealing with professional actors but with professionals who are used to sharing their knowledge. In this case, the script should be a set of topics to cover and their sequence.
Choice of location and wardrobe	The location and wardrobe are important logistical issues, because of the following factors: • The location may be the same as that where they perform the tasks that are to be captured on video. • If the location is not fundamental, we can choose a location with good lighting conditions, with low ambient noise and with backgrounds that do not distract those watching the video. • In the case of wardrobe, this can be important, in the sense of reinforcing the message if: — it is a video of a technical supervisor of the company explaining the functions of a product, it might be more interesting if he is wearing his work uniform; — on the other hand, if it is a director explaining the developments in a project, it would make more sense to have more formal clothing.
Video recording	For the recording of a video, a location with good lighting and no background noise should be chosen. Care should be taken to avoid placing the people in a position of speaking directly into the camera. The easiest way to create a video is the format of an informal conversation. Care should be taken to be slightly turned to one side in relation to the camera, so that expressions and body movements can be seen.

Image 3.06. Example of recording a video, based on a model of informal conversation, using a wireless microphone (see arrow)

This conversation can take place in the format shown in the following figure, that is, face-to-face (seated or standing), with the camera placed over the shoulder of the interviewer and focused on the interviewee, so that the interviewer does not appear in the image.

Image 3.07. Example of recording a video using a model of informal conversation with the camera positioned over the shoulder of the interviewer

Note that the interviewee will tend to speak to the interviewer, thus avoiding the habitual constraint people feel when speaking directly into a camera.

Editing the video

The editing of the video uses the video recordings:

- Cutting out periods of silence and portions having no connection with the content.
- Adding titles, logos and other mandatory references.
- Raising or lowering the volume.
- Reducing background noise.
- Inserting subtitles and call-outs (arrows, text balloons) that facilitate the use of this type of content.

Image 3.08. View of video editing using Windows Live Movie Maker (www.microsoft.com)

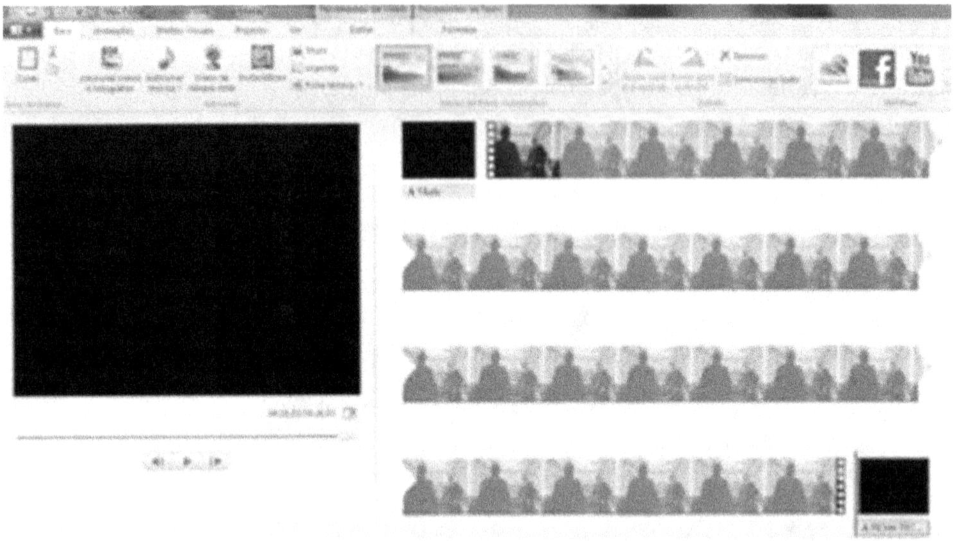

Final result

Depending on the program for editing video that is used, we will have files of different formats:
- WMV – Windows Media Video.
- MP4/FLV/SWF – Flash Outputs.

- MOV – QuickTime Movie.
- AVI – Audio Video Interleave Video File.
- M4V – iPod, iPhone, iTunes Compatible Video.
- RM – RealMedia Streaming Video.
- GIF – Animation File.

Normally, the video editing programs use wizards to help users choose which type of destination is best suited for each video, such as:
- DVD.
- CD.
- Mobile devices, iPod, iPhone, iPad and others.
- YouTube.
- Hard drive.
- Web.

This versatility of formats makes it possible to view the videos in a variety of supports, enabling them to be viewed even when the target audience does not have access to the Internet, because what is important is that the target audience be able to watch this content where and when they need them.

In creating videos, the main precautions to take are:

With the sound:

- To preserve quality when capturing the sound, paying special attention to the ambient sounds, moving away from sources of noise and locations that echo.
- When wireless microphones are used, one must check to see that the range of the receptor installed in the camera is not exceeded and that the camera is not picking up sound from its own microphone.
- To ensure that the microphone is well positioned and does not move whenever the interviewee moves.

With the preparation:

- To test the equipment and be sure there are extra batteries for the equipment being used before beginning the recording.
- To choose a location with good lighting, preferably with natural light. In the event there is no other alternative for the location, a

Advice and precautions

light source should be used, which may be a powerful lamp or even a projector.

- Before beginning the filming, a check should be made to see that that there are no shadows that cover part of the faces of the persons participating in the video.
- From the outset, a check should be made to see if there are elements that would create a distraction, for example:
  — Clothing that is not in line with the goal of the video, such as comic ties, for example.
  — Backgrounds with well-defined images and/or movement, or even live images with passing cars and pedestrians.
  — Presentations or posters that are too visible.

With capturing the image and sound:

- Whenever possible, a tripod should be used to prevent shaking or sudden changes in the plane of view, which cause distractions.
- To ensure a proper diction and that the person speaking into the microphone does not speak too fast.
- To use the widest possible angle of view, in order to present a general view of the non-verbal language of the participants in the video.
- The use of zoom should be kept to an absolute minimum, but it can be very useful when we want to show a detail of a document, piece of equipment, part, etc.
- To check to see if the quality of the sound is acceptable immediately after the first recording, and if so, proceed with the other recordings.

With the editing:

- To reduce to a minimum the cuts during editing, as it is preferable to repeat an entire recording than lose time and work in editing.
- The duration of this type of video should take into consideration the time that the audience will be willing to invest in viewing it.
- The file format should allow for correct viewing on the selected devices.

- There are hundreds of video tutorials on editing programs; viewing some of them may allow you to save time and create higher quality videos.

Applications

Our imagination is the only limitation on the applications that we can make for the use of video in communication, as it is a powerful medium that stands the test of time, in other words, we can make a video and put it online for a certain very time-specific purpose, and years later realize that there are still people who are watching it.

I list a few examples of applications of video in communication:

- Promotion of products, services, works, concepts, ideas, brands, organizations and people.
- Response to solicitations, advantageously substituting traditional e-mail, since it includes non-verbal communication elements.
- Creation of tutorials, training facilitators.
- Sharing of information.
- Demonstration of product functions.
- Recruitment of collaborators, through a video CV or video responses to a series of questions.
- Evaluation of students, where they have to create a video based on the content in a particular subject.
- Creation of your own Web TV channel.

A 2.0 example

Daniel Pereira had a dream of running a half-marathon, but a serious foot problem made that impossible, unless he used a carbon orthopedic prosthesis and running shoes totaling 650 Euros.

He was faced with two options: ask for money from the same old group (family, friends) or give up on the idea.

Daniel chose a third alternative: crowdfunding!

Crowdfunding, or collective financing, is the gaining of capital for initiatives of collective interest through the combination of multiple financing sources, generally individuals interested in the initiative.

Daniel did not write a letter or an e-mail to various potential financers; he made a video and used a platform specializing in crowdfunding – Massivemov.

In his video, Daniel explained to any potential financer his passion for running and what was needed to make his dream come true. In little time, he was able to meet his objectives!

In my opinion, passion attracts money; money does not attract passion, and video is without a doubt a powerful method, natural and appropriate for demonstrating our passion before an increasingly global audience.

Image 3.09. Daniel Pereira's project page on Massivemov (www.massivemov.com)

Live

Until now, we have dealt with previously recorded video, but we live in an extraordinary time when any one of us could have their own Web TV, using previously recorded videos or even making live broadcasts, resorting to platforms such as Livestream (www.livestream.com) or Ustream (www.ustream.tv).

The necessary resources are not many:
• Bandwidth Internet connection.
• A microphone.
• Webcam.

And you are ready to make your first live broadcast to the entire world.

I prefer, when I can, to substitute my webcam with my video camera and a wireless microphone; this gives me better sound quality and I am not reduced to a close up of my face doubled over my computer; I am able to move around, zoom and show detail in other elements.

In this case, it is necessary to use a video card that captures the signal of the camera to the computer; for this, I use a card not much bigger than a low cost thumb drive (under 60 dollars) that allows me total portability.

But it is possible to be even more portable as smartphones and tablets can do live broadcasts, using platforms like Livestream or Ustream, making us all potential creators of live multimedia content.

Image 3.10. Card used for video transmission, via web (www.magix.net)

Television Programs

Framework

I cannot say that I am a person with a lot of television experience; I have been interviewed by television channels now and then all over the world, in the scope of my professional activity and of the books that I wrote.

Possibly, being less experienced allows me to feel closer to those professionals who for some reason or another are invited to make statements in television broadcasts.

In the following points I will give some practical advice and tips on how to deal with live and recorded broadcasts.

Preparation

Imagine that you are working normally and suddenly you receive a phone call from a television station asking for a comment on a subject from your area of knowledge.

, a time is agreed upon, which may follow a pre-set schedule, in the event it is a live broadcast, or at a time to be determined, in the event it is a recorded broadcast or a news piece to be included in a live broadcast.

Regardless of the amount of time you have, there are certain things you can do to be well prepared for the interview:

- Since the journalist in question identified himself, enter his name in Google and see what comes up; be particularly aware of articles and videos, as you will get a clearer idea of his approach.
- Watch previous editions of the program where you will be interviewed, to check audience profile, the way the questions are asked, the usual response time, the depth of the questions, the type of language used, etc.
- Ask the journalist regarding issues such as: most appropriate clothing, how to get to location, reason for contact, if it will be live or recorded, the reason for piece, etc.
- Prepare the message that you intend to communicate.
- When you arrive at the recording location, greet all those involved in the taping in an open and sincere manner.

- Try to have a small informal conversation with the interviewer, in order to create empathy, which will be visible to anyone who later watches the interview and also clarifies the formality to be used in address (first name, sir).
- Look in a mirror and make sure everything is in order; the slightest detail – a crooked tie, an undone button – will be motive for distraction to the audience and panic for you if you discover it mid-broadcast.
- Test the sound and the chair, if you are seated.
- Arrange a way to get the recording of the program; the safest way is to have a DVD recorder programmed for this purpose.

Silence, lights and action!   During

- When given the chance, express your gratitude for the opportunity and greet the audience, but be brief.
- Listen attentively to the questions; if it is a panel or there are questions over the phone, use pen and paper to take notes, writing response topics and the name of the person who asked the question.
- Be aware of the fact that your non-verbal communication is visible and therefore, consider that the cameras are always filming you, even if you think they are not; you will thus avoid unpleasant surprises when you watch the recording.
- Although you are conscious of the existence of cameras, do not look at them directly, speak with your communication partners, leave the camera operators to worry about angles; this way you will feel more relaxed.

What you do after your speech will determine your chances of future opportunities to speak: Afterwards

- Say goodbye to all those involved in recording.
- Place yourself at the journalist's disposal for any additional clarification needed or for future journalistic pieces.
- Evaluate your performance by analyzing the recording; be fair with yourself and learn from your mistakes.

- Look for reactions to your speech on social networks and, if justified, respond.
- Share, on social networks, the recording that was made.

AUDIO COMMUNICATION

Telephone

Framework

There is a funny story about a group of directors of German Post Offices who, at the end of the 19th century, were called into a training session on the advantages of a new device called a telephone. The group thought the initiative was perfectly useless, since the telephone would never be important in communicating between human beings.

How wrong they were! Years later, training sessions appeared on how to use the telephone, sessions which have almost always been directed at workers whose daily tasks depend on phone contact with customers, namely telemarketing services.

Unfortunately, I have noted that, in general terms, the majority of professionals still do not know how to correctly deal with this form of communication, causing conflicts and large losses in productivity.

Preparation

As in all forms of communication, preparation is half the battle on the way to success:

- Clearly define the objective of the phone call (selling, buying, informing, asking for contributions, resolving a potential conflict, etc.).
- On a piece of paper, write all the points that you need to cover during a call, to avoid the famous second call.
- Choose the most appropriate hour to call.
- If necessary and possible, set a contact time with the other person.
- Arrange for the quietest location possible for making the phone call.
- Choose a moment when there are fewer chances of being interrupted by colleagues, customers, etc.

During

A few precautions are necessary during a phone call:

- Call the other person by their name, and often, during the phone call. The word we most enjoy hearing is our name.
- Speak calmly and with good diction.
- Smile as you speak; it may seem stupid, but the people on the other end of the phone know when you are smiling.
- Listen attentively without interrupting, but giving signs of presence with small expressions like: «yes», «right» and «ah!»
- Take notes of what your communication partner is saying and what your response will be.
- Anytime, if it is justified, repeat what the other person said: «If I understood correctly, what you intend...», giving the chance for him to confirm or correct your understanding.
- At the end of the phone call, make a summary of conclusions, clearly defining the next steps.
- Cordially say goodbye.

Afterwards

In a sketch of a Portuguese comedy group, a character says «They talk, talk, but don't say anything»; this is unfortunately the situation with the vast majority of telephone calls in a professional setting, but it does not have to be that way.

To avoid this type of situation, you merely need to send an e-mail with the conclusions of the phone call and what is expected from each party.

But it is also important that you make a self-evaluation of what went well and not so well, with the goal of continued improvement.

Applications

Telephone communications have a very wide range of applications, being especially effective in the following situations:
- Resolving situations of potential conflict.
- When it is convenient to have immediate feedback.
- To accomplish quick brainstorming.
- To avoid in-person meetings.

AUDIO COMMUNICATION

Calling for free?

There are many computer applications that allow us to call for free, or almost free, a good example being VoipBuster, whose main competitor is Skype, since they are competitors in the voice over IP market.

VoipBuster offers the ability to make phone calls to landlines and mobile phones all over the world at much lower prices than Skype does, and in several cases the phone calls are free: for example, when the destination number is a landline in Portugal, North-America or in most European countries (and now, phone calls to mobile phones in the USA are also free!).

The cost of sending an SMS text is also very competitive.

The sound quality of the calls made with this service is not as good as with Skype; so it is best to use a microphone and headphones with an external sound card.

Image 4.01. An inexpensive VoIP international communications service (www.voipbuster.com)

Conference Call

Framework

The conference call allows communication and interaction between people who are located in different places through voice, with the support of phone lines, mobile networks, Internet and VoIP (Voice over IP).

In the case of the Internet, people interact in virtual rooms, which have a predetermined number of participants and a moderator, through audio and text, in real time.

Conference call types

There are basically two ways of making a conference call:

- **Meetme**: the users call an access telephone number and input the room number and password.
- **Dial Out**: in this type, one participant, starting from the conference call, connects with the other participants and brings them into the virtual meeting.

Preparation

Preparing for a conference call follows the same procedures listed for phone calls, but with a few additional elements:

- Designation of a moderator.
- If necessary, when scheduling the conference call, be careful to choose a time that is possible for the participants in different time zones; in these cases it is imperative, when setting the time, to also establish the referencing time zone.
- Preparation of an order of business, agreed upon by the participants.
- Pre-delivery of files and other documents necessary for an informed discussion.
- Definition of the way one should request to speak during the conference call.
- Establishment of a start time and a latest end time.

- If necessary, scheduling a pre-call test of equipment, with some of the participants.
- Each of the participants should arrange for a space that is more or less private to complete their end of the conference call.

During

A conference call is a special conversation, since the participants do not see each other, not permitting the understanding of their non-verbal communication:

- The moderator should assume his role from the beginning, guiding the conversation and managing the time.
- One of the moderator's duties during a conference call is to confirm that all the people are present. There is nothing more unpleasant or awkward than realizing halfway through the call that we are speaking to more people than we originally thought.
- One of the ways to stimulate this type of communication is to previously assign a series of questions, in random order, to each of the participants.
- The participants should be aware that they must wait for the end of the other's contribution before they speak, or the call will easily fall into chaos.
- The participants should say their name before they speak, in case their voices are not well known by the group of participants.
- Although the temptation is great, each one of the participants should cease other simultaneous activities, since the level of information loss will be greater than from doing the same thing during an in-person meeting.
- The use of headsets or Bluetooth earpieces is the most desirable, since it leaves hands free to take notes and type texts to the participants.
- Do not abstain from smiling, or moving your hands and your body. During certain conference calls made with a mobile phone and a Bluetooth set, I even find myself walking through the room, moving my hands, as if there were an audience in front of me. One curious fact: whoever is listening on the other side feels that involvement from the speaker and will also be more focused.

Afterwards	At the end of a conference call, there are certain procedures that should be followed:

• Online availability of the conference call recording to participants and to all interested absentees.

• Delivery of a draft document with the main conclusions.

• Delivery of files presented and other supplemental files or hyperlinks. |
| Applications | Honestly, I do not understand the reasons for the scarce use of this fantastic communication tool.

As an example, mobile providers have allowed for the completion of conference calls between 6 people for over a decade and, in spite of the extremely high level of mobile phone diffusion, this functionality is completely unknown to almost all users.

But, if we knew how to properly use this tool, we could see very significant increases in productivity. Here are a few application examples:

• Using Skype, it is possible to make a conference call involving up to 24 participants, with the possibility of some being on their mobile phones, others on their landlines and others on their Skype accounts, and keeping the conversation fluid, calling and releasing participants during the conference call. The same can be done over VoipBuster, which we focused on earlier.

• We need to quickly hold a meeting with a group of people who are in transit in various locations. With one conference call from our mobile phone, we can reach these people, making a meeting possible without delays due to «traffic».

• We are in a car and cannot stop, since we are already late, but we have to speak right away with «that» client. However, we know it is necessary to take notes or even make some calculations. Solution: support via conference call – we connect with a colleague and ask him or her to take minutes of the meeting and then call the client, starting the conference call by informing that there is a third person listening in to write minutes. At the end of the conference call, that person can send the minutes to the client, for validation. Now that we are talking about it, as you drive, always use a |

hands-free system and never try to make conference calls in urban areas.

- You need to make a conference call between a client and your team. No, you do not need to use expensive conference call systems that place a special type of spaceship in the middle of your table. There are several alternatives, which are cheaper and more flexible:
  - You can use the speakerphone function on a landline or mobile phone.
  - You can use programs such as Skype or VoipBuster to hear the sound through the computer speakers or connect to speakers.

From my own experience, I have found that the conference call is a flexible tool, demanding few resources and reducing the time needed for meetings.

When people participate in conference calls, especially on their mobile phones, they have a sense of the cost of a meeting, reducing their comments to those strictly necessary, and since they do not have an excuse to be late, they arrive on time.

And the costs

There is a notion, quite mistaken, that an in-person meeting is cheaper, since there are no telecommunication costs. Nothing could be more wrong; let's make some calculations.

First, the assumptions of our meeting:

- A meeting lasting 2 hours.
- 5 participants in the meeting.
- Each person has to travel (round-trip) for 1 hour to arrive at the meeting location.
- The average gross salary of each participant is 2000 dollars.
- The cost of each trip is 10 dollars.

Calculation:

Description	Value (in USD)
Cost of 5 people in a meeting for 2 hours	141.41
Cost of commute time of the 5 people	70.70
Cost of commute	50.00
Total	**262.11**

This is a meeting similar to so many others that we hold periodically; however, we do not consider that there are such high direct costs, but this value is merely an estimate, since we did not consider various costs such as:

- Cost of telecommunications used in scheduling the meeting.
- Cost of people who planned meeting logistics.
- Cost of meeting location.
- And, likely, the highest invisible cost – the cost of opportunity. In participating in a meeting, these people are not doing billable work for the company.
- Normally, there are delays at the beginning and a temptation to socialize in the meeting, leading to more dead times in an in-person meeting and a tendency to schedule new meetings.

After making these calculations at your organization, you will have a weighty argument to reduce the number of useless meetings and use alternatives such as conference calls or video conferences.

Please do not conclude that I am against in-person meetings; on the contrary, they are very important and valued, but what I am saying is that we should be critical in choosing which communication method to use.

Podcast

Framework

If you feel more comfortable speaking than writing, make your voice heard all over the world, make a podcast!

What is a podcast?

The term podcast emerged in 2004, by combining the words iPod (MP4 reader) and broadcast (radio or television transmission). The author of a podcast is called a podcaster.

A podcast is a set of audio files which are published one time or periodically and are capable of being downloaded from a site or a podcatcher (iTunes, Zune, Juice or Wimap).

Such programs make it possible to download these media files to the computer or to MP3 and MP4 readers, allowing one to listen to them offline, which is particularly useful while travelling or in locations where access to Internet is limited.

Most likely, due to the habit of using audio books in the USA, podcasts are more popular there than in Europe.

Making a podcast

However, this is not a reason to not use podcasts to make our work, ideas or desires known. The important thing to determine is if your target market will listen, and the only sure way to find out is to try.

Using a site like MyPodcast and a computer with a microphone, you can:

- Use the client program, provided by the site, which allows you to create the podcast and upload it to the MyPodcast site in a type of audio blog.
- Keep in mind that the visitors to your MyPodcast page can subscribe to your podcasts and receive them, as they are published, through iTunes, Google Reader, MyYahoo or through other podcasters.

Try and you will see that in 5 minutes you will create your audio blog.

Image 4.02. MyPodcast allows you to create and distribute podcasts (www.mypodcast.com)

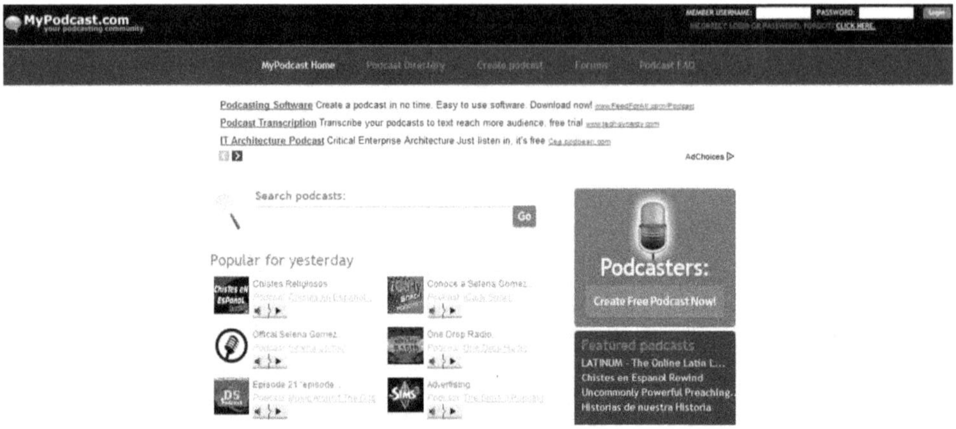

Equipment

Several types of equipment can be used to produce podcasts:

- Microphone, connected to the computer with or without an external sound card. The advantage of a microphone with an incorporated sound card is that it does not depend on the computer sound card, which in many cases does not have the sufficient quality to record sound through simple microphones.
- Computer for audio editing.
- Portable mp3 recorder, such as an iPod Nano.

Image 4.03. Example of a microphone with a sound card and USB connection (www.plantronics.com)

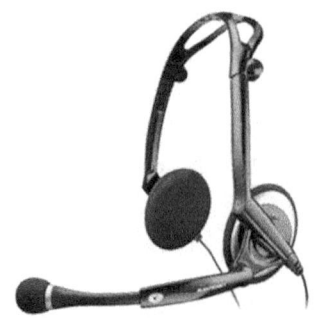

Software

Besides the unique applications of podcast platforms, such as MyPodcast, there is a great variety of software available for editing audio. However, for reasons of simplicity, we will cover two programs:

- **Audacity** (http://audacity.sourceforge.net) – a free program with many editing possibilities and which is perfect for the user with little audio editing knowledge.
- **Camtasia Studio** (www.techsmith.com/camtasia) – a video editing program (not free) that also allows audio editing.

Creation process

The podcast creation process will go through the following phases:
1. Creating script.
2. Recording content.
3. Editing audio.
4. Final result.

Creating the script

There are two possibilities for creating the script:
- In the case of there being perfectly identified steps, it would be best to create a detailed and previously validated script, to be read and recorded later.
- If it is based on an interview with the person at his workplace, it would be best to merely clearly identify the podcast objective.

Recording audio content

The type of recording will depend on the option chosen in the previous phase:
- Should the recording be made using a previously written document, a computer with a good microphone or any other portable mp3 recording device can be used.
- In the case of the recording being made at another location, it may be more convenient to use a portable recording device to gather audio from the person we intend to record.

| Editing audio | In editing the audio, audio editing programs such as those mentioned in the previous section will be used. The main tasks to perform in this phase will be:

- To eliminate periods of silence at the start, in the middle and at the end.
- To eliminate unnecessary content.
- To raise or lower the volume when necessary. |

Image 4.04. View of audio editing using Audacity (http://audacity.sourceforge.net)

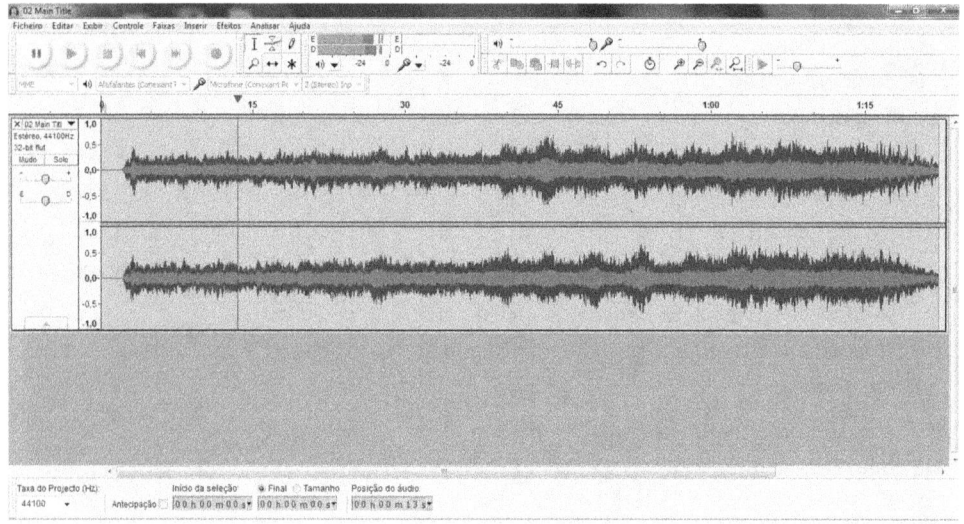

| Final result | As a result of the foregoing phases, we will have podcasts in audio files that can assume various formats, mp3 being the most common, due to its ease of use in multiple supports. |

| Advice and precautions | In creating podcasts in audio format, the main precautions to take should be:

- To preserve quality when capturing the sound, paying special attention to the surrounding sounds, moving away from sources of noise. |

- To test the equipment and be sure you have extra batteries before beginning the recording.
- To ensure a proper diction and that the person speaking into the microphone does not speak too fast.
- To check to see if the quality of the sound is acceptable immediately after the first recording, and if so, proceed with the other recordings.
- To reduce to a minimum the cuts in the editing; it is preferable to repeat the whole recording than to leave it for the editing phase.
- To keep the length of a podcast to no more than 3 minutes to avoid breaks in concentration. There are platforms, such as Blaving, that do not allow more than 2 minutes, in keeping with its description as an «audio version of Twitter».

Image 4.05. Blaving, an audio message sharing site (www.blaving.com)

Creating podcasts can serve various purposes: Applications
- Sharing ideas and thoughts in a quick and flexible manner.
- Informing customers and other stakeholders.

- Providing sections of content from articles, books or trainings, as a way to attract listeners to purchase the remainder.
- Creating your own radio program.

Radio

Framework	Radio is still a powerful medium of live communication, and increasingly pre-recorded communication, due to many programs being available in podcast format. A professional must be ready to communicate through this medium. I will share with you what I have learned in dozens of interviews and participations in radio programs.
Preparation	Preparing for a radio program is very similar, with slight variations, to what we previously saw for a television program: • Analyze the journalist and program profile. Many radio programs have their own Facebook page for interactions with the listeners. • Listen to previous editions of the program where you will be interviewed, to verify the audience profile, the way the questions are asked, the usual response time, the depth of the questions asked, the type of language used, etc. • Carefully think out what you would like to say, what points to cover, in short, what message you want to give, creating small messages with impact and easily understood – the so-called «sound-bites». • In arriving at the recording location, greet all those involved with the broadcast, in an open and sincere way. • Seek to have a small informal conversation with the interviewer, in order to create empathy that will be perceptible to anyone listening to the interview and also to establish forms of address. • Test the sound, especially the correct distance from the microphone, since microphones are attached to the tables at radio stations, but may still be adjusted.

- Arrange for a way to obtain a recording of the program; currently the easiest way is to check and see if there is a podcast version of the program.

And the light comes on «On the Air»:   *During*

- When you are given opportunity, express gratitude for the opportunity and greet the audience, but be brief.
- Listen attentively to the questions; if it is a panel or there are questions over the phone, use paper and pen to take notes, indicating response topics and the name of the person asking the question.
- Although your non-verbal communication is not visible, do not remain paralyzed; move around, use hand gestures, but be careful with the microphone.
- Use the headset only if you need to, for example when you need to hear the questions posed by listeners or from other studios. For those not as used to it, hearing one's own voice on the headphones with delay can cause considerable mental confusion.
- Without losing your own style, follow the interviewer's style.

Just as we saw for television programs, what you do after your contribution will determine any future opportunities: *Afterwards*

- Say goodbye to all those involved in the recording.
- Place yourself at the journalist's disposal for any additional clarification needed or for future journalistic pieces.
- Evaluate your performance by analyzing the recording; be fair with yourself and learn from your mistakes.
- Look for reactions to your speech on social networks, especially the program's pages, and if justified, respond.
- Share, on social networks, the hyperlink to the podcast of the program.

5
WRITTEN COMMUNICATION

E-mail

Framework

If we were to investigate and look for the reasons that caused a certain conflict between people or organizations, we would find many times that a written document was the object of diverse interpretations, which, most likely, never occurred to the one who wrote it.

Writing reduces communication, has ambiguities, interpretations and, more importantly, is void of non-verbal communication, which means that more than 90% of what you intend to communicate is lost.

What I will share next is my experience in the use of this extraordinary communication tool that is the e-mail, but that is used so poorly that it enslaves its users.

80% of e-mail is useless

We feel daily, especially through e-mail, a growing pressure to communicate more and more in writing, as it is necessary to have proof of everything we do.

I think that more than 80% of e-mail messages merely create noise, and I will give some significant examples below of this brutal loss of time, with multiplying effects.

The e-mail in Aramaic

In 2010, I had the opportunity to visit Malula, in Syria, location where the residents seek to maintain the tradition of speaking Aramaic, which was the language that Jesus Christ spoke.

When I heard this language lost in time, I did not understand anything, which is obvious, much like so many e-mail messages I receive.

Whoever wrote them was probably multitasking, thinking that he was capable of writing that message at the same time he answered the phone and made signals to the colleague in front.

And the results are often disastrous –incomplete messages at the least; at their worst, interpretations completely contrary to what was intended.

Forget grammar school rules

In any elementary education system, the students are prepared to demonstrate richness in vocabulary, use figures of speech, follow punctuation rules, etc., all with the ultimate goal of writing like novelists, poets and authors.

The problem is that 99.999% of these students will never be writers, but all of them will use written language to communicate with other human beings.

So, when we are speaking of business writing, we need to forget some grade school rules and be more pragmatic, or we risk being ineffective.

Here are a few business writing rules:

- Forget the thesaurus – there is no problem in repeating the same terms, if we are referring to the same thing. An example: If I say that A is equal to B and further ahead in the text I say B is identical to C (to not say it again in the same way), I am placing doubt in the reader's mind.

- Paragraphs should not be more than 3 to 5 lines long – in school we learned that we could have a paragraph with 16 lines and, to facilitate, divide it in sentences. The problem is that nobody can understand technical text with paragraphs longer than 5 lines without having to reread it several times. Try dividing an e-mail with extensive paragraphs into smaller ones of 3 to 5 lines. You will see that suddenly everything becomes clearer, especially when you must read it on a computer screen, tablet or smartphone.

- First, the conclusion, then the details – we have always learned that all texts have three parts: introduction, development and conclusion. But an e-mail is a type of communication that we want to be fast, and so, from the reader's point of view, he wants to act depending on what the sender's conclusions are, and only in case of doubt is he interested in the process of reaching those conclusions. Therefore, in business terms we have: conclusion first, and then the details, where we include hyperlinks, images or attached documents.

- Labeling blocks of text – with the daily flood of e-mails, we are tempted to do so-called «diagonal reading», which many times is responsible for the creation of so many more e-mails, phone calls and useless meetings. If we all know that everyone does it, why not change the way we write? If we divide our e-mails in various blocks and each block has at the top a caption (conclusion, proposed action, scenarios, etc.), we help everyone to navigate our e-mails.

There are too many e-mails in this world and I am not just speaking of pure e-mail, I include here written messages of all types, even the famous instant messaging.

Because it is so easy to send and receive written electronic messages without the need to use heavy paper files, writing has increased, without increasing the quality and much less the efficiency.

E-mail should not be the first option when we need immediate feedback from an interlocutor. There are people who resolve that situation, using e-mail as if it were chat; personally, I think that is the worst thing we could do.

When there is a need for immediate feedback, we should speak in person, by phone, Skype or video conference, and at the end create a message with our conclusions from the conversation, since in the majority of cases it is not the process, but the conclusions reached that matter.

In the same way, when people have to write they tend to be less sincere and more defensive, with the opposite behavior occurring when the contact is made out loud.

A conversation is worth more than a thousand e-mails

The toilet-paper e-mail is one of the biggest problems in the use of e-mail in any organization, with their creators normally thinking that they are proceeding as correctly as possible, when in fact they are clogging communication channels with _____ (fill in the blank as you wish).

A typical example:

- John sends Joanna an e-mail message about a certain subject.
- Joanna receives it, but as she daily receives hundreds of messages, she will respond two to three days later.
- John receives Joanna's response, but as he also receives hundreds of messages daily, he will delay the same amount of time in seeing the message.

The toilet paper type of e-mail

- When John goes to handle the e-mail response from Joanna, sufficient days have passed to not remember exactly what he initially wanted, so that he may send a response that is not entirely coherent with his first message.
- And this ping-pong continues for more days, weeks or even months.

If everyone knows that this type of situation happens, why does it repeat itself?

The reason is simple: it is a waste of time in easy installments where neither party totals up time lost.

We have here another case where live and in-person communication, followed by a summary e-mail, can avoid the toilet-paper e-mail.

Here goes a cc for all

Long ago, the cc (carbon copy) almost did not exist, since it cost a lot of money and effort to do it for letters or even faxes, but now that the cost and the effort are non-existent, here we go, even sending a cc to the cat.

With this type of attitude, we pollute e-mail inboxes, leading recipients to start filtering messages received and, since that filtering is not perfect, important messages get deleted.

Additionally, recipients who respond to all will start another mountain of e-mails and even create situations of conflict. Some recipients will make comments that are not well founded, since they read a gigantic e-mail on a diagonal – result: chaos.

So, it is important, in certain situations where a message is sent to various people but no group discussion is intended, to place all the addresses in the bcc field (blind carbon copy). In this way, each one receives a message, without knowing who else received it.

The 13th version e-mail

A significant portion of e-mails sent represents the mailing of files and all their subsequent versions, to which we add messages asking which version is correct and the corresponding new versions.

Until not long ago, it was difficult for individuals and small organizations to solve this problem, but now there is no excuse for it. With tools like Dropbox or Google docs, sharing files is easier, furthermore, editing them can send automatic notifications to all those interested.

The use of these types of tools, besides simplifying processes of management and creation of information, has a collateral effect of eliminating many useless e-mail messages.

The big fear of someone who sends an e-mail message is that it will not reach its destination. There are good grounds for this fear, since there are many e-mails that get lost for various reasons.

Of course we can request a read receipt on all our messages, but even so we will have problems, since there are people who have opted to never respond to read receipts, or who plainly and simply do not respond.

It is good practice to request read receipts and to respond to requests for read receipts, but this is not sufficient; in relationships with some level of permanency (customers, co-workers, suppliers, etc.), it may be interesting to create «virtual rooms» through collaborative platforms, where even if a message is lost in e-mail, anyone can go to the corresponding room and see all the messages, files and hyperlinks that were exchanged.

My experience is that with some interlocutors, it is easier to communicate using messages on applications such as Facebook, Linkedin or Twitter.

The e-mail confirming the receipt of an e-mail

It has never been so easy scheduling meetings, but the customary scheduling of meetings happens through an avalanche of e-mails in all directions until a consensual date and time are arrived at, which somebody has to discover.

Because of these situations, somebody decided to invent a web tool that functions well for scheduling meetings. Using Doodle reduces the need to write and read many useless e-mails.

Scheduling meetings through an avalanche of e-mails

Image 5.01. Doodle: a web tool for scheduling meetings (www.doodle.com)

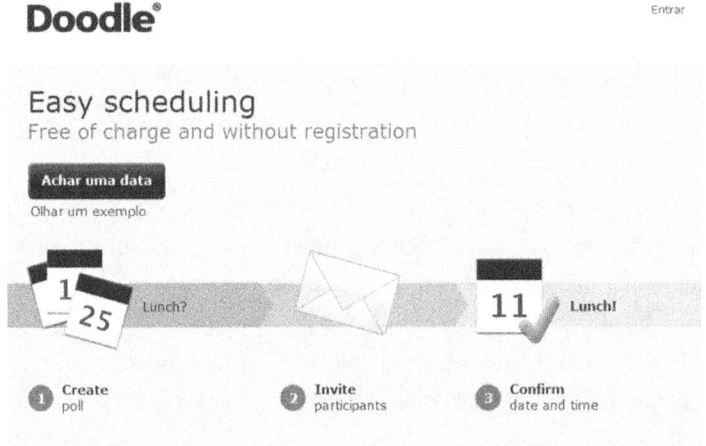

The useless information e-mail

Imagine a situation where you need to ask for several different pieces of information from different suppliers to make a purchasing decision.

What you normally do is send an e-mail to the suppliers and then each of them responds to your message in a different way, that is:

- They will sequence differently each of the pieces of information requested.
- They will accidently forget to include certain information.
- At some point, they will lose some of the messages.

Because these situations actually happen, you will lose more time with new e-mails, phone calls or even meetings. And since you are an organized person, in the end you will compile all the information you were able to gather in a table to make the proposals easier to compare.

If you use a Google form, you can reduce the time it takes to accomplish this simple task by up 90%.

How do you do that? Easy:

1. First, create a form in Google docs; you will see that you can create a very professional form in less than 5 minutes.

2. Send an e-mail to all suppliers with the hyperlink to the form or you can place it directly on your site.
3. Each supplier fills out the form and in submitting it will receive a message confirming it has been received.
4. Even if a supplier calls you (because the Internet is down), you can insert in the form the information that the supplier is providing.
5. In the Google form, you will see the result in two formats: graphics or spread sheet, just as in the traditional way, but with an innovation that may be very interesting: all entries have a timestamp that indicates time and date that the supplier really submitted his proposal, eliminating many excuses.

This example can be extrapolated into a myriad of situations in which we need to compile, quickly and effectively, information from various sources.

Image 5.02. Google form: an available tool from Google docs (https://docs.google.com)

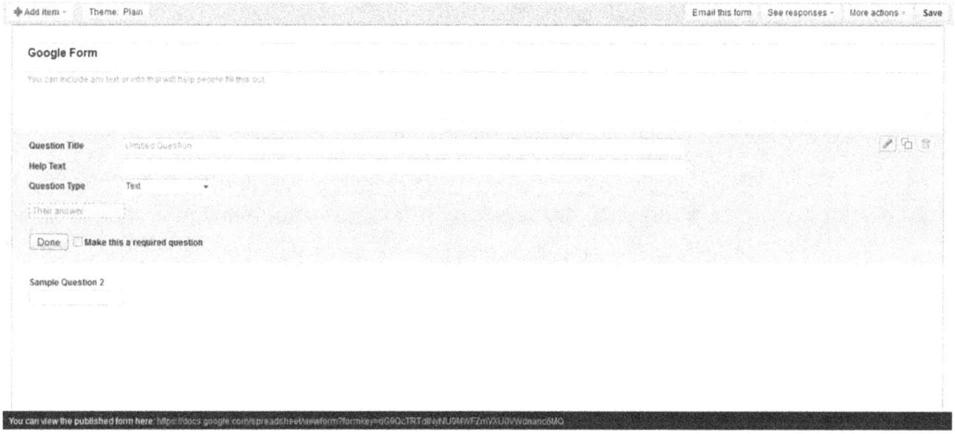

Now, I love my smartphone and I love reading e-mails on it, but attention! I see a dangerous tendency, capable of creating errors and even conflicts, due to the use of the smartphone as the primary tool in responding to e-mails.

How cool is my smartphone?!

The danger of this type of action can be described in the following way:

- The messages sent are often incomplete, since it is uncomfortable to read long messages.
- Reading e-mails is done even more on a diagonal, leaving room for poor interpretations.

This type of device is great for:

- Following up on situations in progress.
- Responding with brief messages, in situations calling for brief messages.

So, be smart in using your smartphone, because you may at times think that you are gaining time, when in fact you are losing it due to the problems you could create.

Very polite, not very assertive

I hate e-mails asking for something which end with:

- Urgent.
- As soon as possible.
- Due yesterday.
- With the greatest speed.
- And for those who like abbreviations: ASAP.

The problem with these expressions is that they do not clearly define a due date, leaving it up to the free interpretation of the reader to determine the date and time of delivery.

It is a very polite approach, but not assertive at all, and you will have as a result new e-mails, phone calls and even meetings to, yes, now establish deadlines in a way that we could already call pre-conflict or even conflict.

When I receive e-mails with these types of expressions, I immediately reply asking for the deadline and, when I receive it, I say whether or not I can commit to the request.

And you can be sure that several things will happen:

- The sender gets the impression that I am a strict and organized person, which is a good thing.

- At the end of one or two similar situations, the sender will always start including the deadline, since he knows that he will have to respond to another e-mail if he does not.
- If I am not able to complete the request in the timeframe given, the sender has several options:
  — Retract the request.
  — Extend the deadline.
  — Find somebody who is able to meet the deadline.
- This way, we will not disillusion the sender, creating unrealistic expectations.

I would like to end this section about e-mail communication with a golden rule of e-mail use: «All e-mails will be forwarded».

All e-mails will be forwarded

This means that:

- You should not place two completely distinct subjects in one e-mail. For example: sending an e-mail with information regarding a product and take advantage of that same message to ask the colleague if he is going to watch the game at his house, since it could be awkward if the boss ends up receiving this e-mail.
- We should not make comments that we do not want others to see. Example: sending a colleague a list of prices requested by a client and adding a funny comment about that same client. Imagine if your colleague, by mistake, forwards the message to the client, forgetting to remove the comment – fired for sure!

Reports and Proposals

Reports and proposals are necessary documents in any organization; however, those who write them have rarely had the specific training to do so, which means that learning is made through trial and error, at the cost of the organization itself or the person's own professional career.

Framework

| Most common mistakes | In my professional experience, I have had to read a substantial amount of these types of documents and I share here the most common errors I have encountered:
- No objective.
- Lack of organization.
- Attempt to make it a literary piece.
- Spelling mistakes. |

| Preparation | Some preparation is necessary in creating this type of document, such as:
- Clearly defining the document's objective.
- Identifying the recipients, especially their needs and their knowledge on the subject.
- Sources to be used.
- Information needed for creating the document.
- If necessary, verifying if there are any royalties on certain content to be included. |

| Organization of topics | One of the most common errors is trying to start to write, in the hope that ideas will magically start flowing from the fingers.

Of course this only happens on rare occasions, so we have the initial mental blocks or very unbalanced documents, which were noticeably worked on with great detail in the beginning, but in the end were rushed since deadlines were looming. The resulting content is almost a telegram.

The way to keep this from happening is to invest time in organizing the document structure – only after having a perfectly defined structure should we start to create the document.

A tool that I always recommend is MindMapping (already referred to in Chapter 2), which allows me to structure a document, no matter how complex it may be, on a simple piece of paper, allowing me to visualize the document as a whole, avoiding omissions and duplications. |

Normally, I work with free software in creating mind maps – Freemind. More information about this application is available at http://freemind.sourceforge.net/.

In this type of document, we should also forget the grade school rules we looked at writing e-mails.

Writing should be done thinking about the reader and not the writer; so, we speak first of identifying the objective and the recipients, in order to be more effective.

In the creation phase, I use a writing method called «Technical Writing»; since it is so flexible, I use it in all the documents I create, namely this book.

Writing

The Technical Writing method is based on seven principles:

- **Principle of Division** – the information should be divided into small and manageable units.
- **Principle of Relevance** – you should make sure that all information in a unit refers to a main point, based on the objective or the function of that information for the reader. In other words:
  — similar subjects should be grouped;
  — those not related should be excluded from each unit.
- **Principle of Captioning** – after organizing phrases about the same subject in manageable units, you should create a caption in each unit of information.
- **Principle of Coherence** – for similar subjects, you should use similar words, captions, formats, organizations and sequences.
- **Principle of Integrated Graphics** – you should use diagrams, tables, images, etc. as integral parts of the text and not as something added on after the writing is complete. In a document made following this method, the graphics accompany the text; for example, no reference is made on page 80 to the graphic on page 32; if necessary, repeat the graphic so that the reader has the graphic and its corresponding analysis on the same page.
- **Principle of Accessibility to Details** – you should write with a level of detail that makes the information immediately accessible

Principles of the Technical Writing method

to the reader and the document useful to all professionals. In other words, the information the reader needs should be where he needs it.

- **The Hierarchy of Principles of Division and Captioning** – it is necessary to organize small and relevantly hierarchical units of information and provide caption(s) to the bigger group(s) they created. As the set of topics increases, beyond the limit established by the beginning of the division, the reader starts to have difficulty understanding or remembering the information. Research indicates that people understand and memorize larger quantities of information better if it is organized in groups of 5 to 9 topics each.

Image 5.03. The Technical Writing method has been developed by Information Mapping (www.infomap.com)

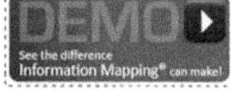

Articles

Framework

We will cover here the writing of articles for any type of media – newspapers, magazines, sites, blogs, annual editions, etc. on subjects in which we have specific training or experience to share.

Prior questions

When we are asked to write an article we should, to be effective, request the following information:

- Delivery deadline – if we are not able to deliver in the time requested, it is better to inform immediately than create a situation of lack of content on the due date.
- Minimum and maximum number of characters (with or without spaces) – this question is often forgotten, leading to later increasing the content or making cuts, potentially completely altering the original text.
- Readers' characteristics – in terms of age, knowledge of the subject to be covered, profession and hierarchical and socio-cultural level.
- Clearly establish the desired content and which register of language to use (more or less formal).
- Find out whether there is a need to supply images, graphics or photographs.

Preparation

After having the answers to the listed questions, we should prepare for the creation of the article.

- Compile texts and other relevant information.
- Gather images we will use, confirming if there are related royalties.
- Define the topics to be covered and only then their order. At this point, I recommend also using a mind map, which will allow you to easily create topics and subtopics, effortlessly rearranging them.

- Clearly define when you will start and end writing the article. Never define the delivery due date as the completion due date. You could run into a setback and your reputation would be on the line.

Practical advice

Some important issues worthy of considering when writing an article:

- Unless it is a scientific article, be simple in your approach, in a way that any person can read it and understand it.
- Write thinking of your audience and not yourself.
- Whenever possible, dialogue with the readers, to better understand the impact you are or are not having.
- Gather periodic feedback from the editor.

There is life after publishing

The life of an article does not necessarily end with its publication; it could merely be its start:

- Try to quickly receive the article in hard copy or digital format (hyperlink or file, normally as pdf).
- If the article is available online and has sharing buttons to social networks, share it quickly.
- Should the article not be available online, you can, using the file or digital version, share it through your site onto the social networks.
- You can even use a site like Issuu (www.issuu.com) to not only make the article available online as an e-book, but to also share it on social networks.

These types of initiatives allow us to considerably increase our reader base.

Image 5.04. Issuu, a tool for publishing articles and books on the web in e-book format (www.issuu.com)

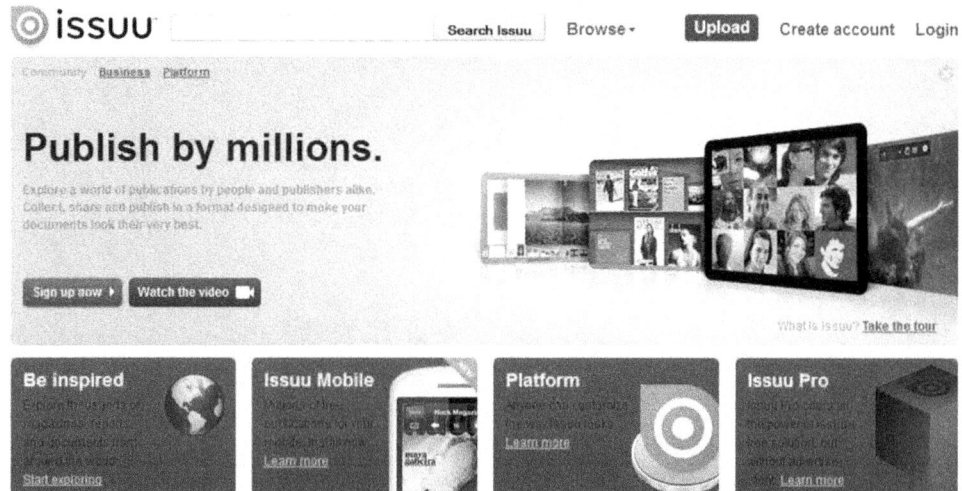

6

PRACTICAL ADVICE

Preparation

Framework

Sun Tzu once said: «Every battle is won before it is ever fought». Preparing for communication is the foundation for success since it gives self-confidence to whoever is communicating and prepares him for succeeding in any type of situation.

In this section we will see what you should take into consideration during the preparation phase for communication.

Promotion

Communicating implies the existence of recipients of that communication, and we should therefore create the conditions to have a well set-up room of recipients, whether that room is physical or virtual.

In promoting, we can use two different types of means:

- **Offline:**
  — Conferences and seminars, announcing upcoming communication opportunities.
  — On the back of business cards.
  — Brochures.
  — Roll-ups and posters in the room.

- **Online:**
  — Podcasting.
  — Vodcasting.
  — Social networks (Facebook, Linkedin, Twitter, YouTube, etc.).
  — Blog.
  — Website.

Time management

Time management is always a great concern for whoever needs to communicate, but it is also true that this concern is greater during the actual communication than during its preparation.

Time management for any communication depends on how it was prepared, because of that we must:

- Establish times for each content section to be presented and for the activities planned. But we should not fall into the mistake of dividing the number of slides by the time available, since one slide could take 15 minutes and a set of 10 slides could be presented in 30 seconds.

- Set aside time for questions and answers.
- If necessary, prepare different presentation situations. It has happened to me, because of different circumstances, to be asked to reduce my presentation time to a half or even a tenth.
- Arriving at the presentation location on time, with a margin for any unforeseen circumstance, is a great start.
- Have a clock in the room that is easy to see. This is an important matter. In Mediterranean countries, in Latin America and in Africa, it is rare to find a clock at the speaker's disposal in an auditorium and even rarer in a plain meeting room. In countries in Northern Europe, USA and Japan, the clock is an indispensable element in any room. In very large auditoriums, request a clock or take a large one, because the speaker who keeps looking at his wrist watch, besides losing concentration, does not demonstrate great confidence to his audience.
- Look at the room, to verify entrance and exit times, time for changing formats and the need for assistants in distributing or gathering items from the audience.

Visualization

It would be ideal to have a general rehearsal of the presentation, but in the overwhelming number of cases, this will not be possible.

For this reason, visualizing the communication could be the best way to verify timing and dynamics, ensuring adequate preparation for any type of communication.

Self-motivation

As we have seen, whoever communicates, transfers emotions to the audience, that is, if he is confident, motivated and passionate about the subject, the audience reacts positively, transmitting that feeling to the communicator, giving him even further confidence, and so on, in a positive spiral.

But if he is nervous, insecure, tired and unmotivated, the audience will react negatively, transmitting that feeling to the communicator, creating more and more problems for him, in a negative spiral.

One day you might be forced to give a presentation at a time when you are not in the least motivated to do so but, since you have to do it, you quickly rush to get the matter over with, which almost always has disastrous results.

My suggestion is to follow four simple steps, even before you start:
- Isolate yourself (it could be in the bathroom).
- Straighten your back.
- Lift your head, stretching it towards the ceiling.
- Give the biggest smile your mouth will allow.

Thanks to this authentic injection of endorphins, you are ready to face any audience. You have to try it to believe it!

One of the biggest sources of stress for any communicator is to realize just before or during the communication that he has forgotten a somewhat essential element.

Checklist

So, it is important to always have a checklist, depending on the situation of each communicator. Here is mine:
- Promotional material.
- Video projector.
- Evaluation questionnaires.
- Computer.
- Business cards.
- Audio cable.
- Speakers.
- Wireless mouse.
- Camera.
- Books.
- Video camera.
- Microphone.
- Tripod.
- Adhesive tape.
- Bell.
- Watch.

And do not forget to wear comfortable clothing, since that makes all the difference in the quality of your presentation.

During

Framework

As we saw earlier, during communication we have three very distinct phases:
- Opening.
- Content.
- Closing.

Here are some pieces of advice that cross-over into any one of these phases.

Empathy

Any successful communicator establishes a high level of empathy with his audience. To better understand the importance of empathy, I will give you an example.

If you tune a radio station to its exact frequency, for example 89.50 MHz, you will have great sound. If you tune it to 89.55 or 89.45 MHz, you will hear the same station, but now with more static, which will make listening to it difficult.

Empathy means identifying yourself with your audience and your audience identifying itself with you, creating a communication channel free of static; for that reason, it is very important to see the presentation through our audience's eyes.

Questions

In a talk, if there is a possibility for the audience to ask questions, we can say that there are two times for that to happen:

- During the presentation – it increases interactivity, but on the other hand, questions may affect the rhythm of communication.
- At the end of the presentation – this allows for clearly defining a time for presentation and a time for questions, but on the other hand, a part of the audience will give up asking relevant questions, feeling that the moment has passed.

The decision regarding the timing for questions will depend on your priorities as a communicator.

- If you encourage interactivity, the questions should be asked during the presentation.
- If you need to present complex ideas without interruptions, in order to do so in the time allotted, then the questions should be asked at the end.

Interactivity

Audiences are less and less open to being treated as mere recipients of knowledge from a communicator; they want to be part of the process and therefore they appreciate some interactivity.

It is my experience that even the shortest speech can have a certain dose of interactivity, but I also see some communications that, when squeezed to extract their content, have nothing concrete to show, since they were merely a set of animated activities.

Interactivity in a speech is not an end in and of itself; it is one of the methods at the disposal of any communicator.

Humor

In the majority of circumstances, using humor is appropriate, since it is a way of captivating an audience and keeping its attention during a message.

A few precautions when using humor:

- Using humor in the communication about serious topics is counterproductive.
- Be careful with taboos; making jokes about matters like politics, religion, gender and race can cost you dearly.
- Using humor about your own self is normally very safe.
- Beware: a communicator is not a comedian, but a comedian can be a great communicator.
- The moderate use of humor is a communicator's tool; excessive use can transform the communicator into a fool.

Multitasking

We live in a world where we are pressured to multitask, something that really does not exist, as we saw before. What does exist is switchtasking, that is, jumping from activity to activity doing them poorly, but with the feeling that much has been done.

When communicating in any method to any type of audience, forget everything around you and concentrate on what is most important, your audience's needs and answers (verbal and non-verbal).

Feedback

Framework	An essential part of communication is feedback; without feedback there is no communication. For this reason, the communicator should seek to gain feedback from his audiences, doing so, in person or online.
Feedback phases	To gather quality feedback, we should follow four steps: • Receive. • Appreciate. • Summarize. • Ask.
Receive	When we are receiving feedback, we should use active listening, listen until the end and not jump to conclusions.
Appreciate	Whoever gave us feedback, positive or negative, went to the effort to do so and, as such, is contributing to our development as communicators. We should, therefore, appreciate this effort and thank them.
Summarize	To be sure that we understood all the points of the feedback given, we should seek to summarize the main points and confirm with the person who gave us the feedback that we understand the idea that they were trying to communicate.

Having somebody who is available to give us feedback on our communication is a golden opportunity. We should take advantage of this by asking questions regarding results, interaction, language used, etc., in order to have the most complete picture possible of the impact of our communication.

Ask

In May 2011, I was present for the 8th time as speaker at an international conference of the American Society for Training and Development (www.astd.org). As usual, this conference had almost 10,000 participants, divided into a hundreds of parallel sessions.

Example of online feedback

I always counted on a good system of session evaluations, based on paper questionnaires distributed during each session, gathered laboriously by a small army of volunteers and made available for speakers' reviews in a breakout room.

This time, the use of Twitter was encouraged, with hashtags for the conference itself (#astd2011) and a code for each session. During the sessions, participants – through smartphones, tablets and computers – made comments, quoting, criticizing and complementing the speakers.

I found several cases of bloggers and journalists who, instead of using an old notebook to prepare their articles, used Twitter as a simple and fast way of compiling the most relevant ideas presented.

In my case, it was interesting to see the positive evaluation of my session given by those present, even before they left the room. There were even people who during the session had made a «retweet» (resending messages placed by others on Twitter), expanding my audience.

After analyzing the messages placed on Twitter regarding my session, I did several things:

- I thanked those who used flattering words; fortunately, there were no negative comments.
- I sent more content to those who seemed to need it or who were more active.
- I started following on Twitter a few of the participants who made comments, since we had common interests.

This is a good example of a fuzzy border between online and offline. There is a fusion of these two realities, which until very recently were very easy to separate.

Going Further

Framework	When I turned 18, I got my driving license and considered myself skilled enough to drive. My father shattered my illusions, telling me that I would only know how to drive after I had driven more than 50,000 km. The truth is he was right.

With this book I intend to help people communicate better, regardless of the method used, but practice and going beyond this book is what will help them go even further in the art of communicating in the 21st century. |
| Seeing the best | Today, it is very easy to learn from the best communicators in the world. We no longer have to run after them; they enter our homes through television, radio, newspapers and the Internet.

I strongly advise watching the videos from conferences such as TED (www.ted.com), Ignite (www.igniteshow.com), PopTech (www.poptech.org) and so many others available on applications such as YouTube.

A word of caution, though: be inspired by these communicators, but never try to imitate! You will sound phony, losing all credibility with your audience. |
| Volunteering | In my college days, I learned that just going to class is not enough to be well prepared for a career. So, I worked a lot (perhaps more than on my studies) in organizations such as AIESEC, Association of Students, and in the Assembly of Representatives of my college (ISEG).

All these jobs were volunteer work, without any type of remuneration, but essential to my professional career. In any of these organizations, |

I had to give dozens of presentations, sell my ideas and my organization, etc.

I learned that volunteering is a phenomenal way of training our skills and improving them. Therefore, when I left college, I continued to volunteer at various organizations and haven't stopped yet. We can always be better.

I highlight three international non-profit organizations that can help anyone be a better communicator through experiences relevant to communication:
- AIESEC.
- JCI.
- Toastmasters.

AIESEC (originally, an acronym for Association Internationale des Etudiants en Sciences Economiques et Commerciales, in English: International Association of Students in Economic and Business Sciences) is an international group that works with exchange students in the 110 countries where it is located.

AIESEC

Today, the full name is no longer used, since its members belong to other areas of study, and only the acronym AIESEC is used.

I pride myself in being alumni of AIESEC ISEG (Portugal). The work I did at the end of the 80s was of great importance to my professional future.

And so, I advise all university students to not settle for the academic experience of the university, and to go further still before entering the workforce.

They should participate in their student associations, in AIESEC, JADE (www.jadeportugal.org), or ELSA (www.elsaportugal.org) and the many other associations of national and international caliber, which can be of extraordinary importance in the development of new skills, primarily in the area of communicating.

Image 6.01. Official site of AIESEC (www.aiesec.org)

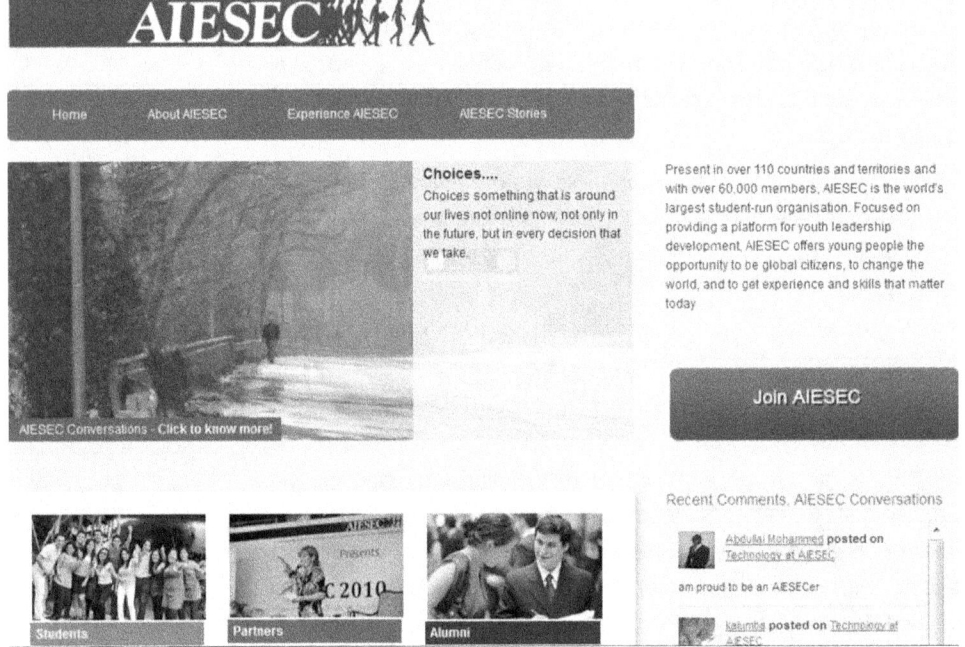

JCI

JCI – Junior Chamber International – is a global network of active young citizens, with approximately 200 thousand members distributed over 120 countries. It is an organization where leaders and future leaders gather, learn and grow.

Its members are leaders, entrepreneurial, creative and bold, between 18 and 40 years old. JCI is formed by the union of students and professionals in all areas of expertise.

When I left college, I felt the need for an association like AIESEC where I could continue to improve my skills and give something to the community. Finally, in 2004, I was one of the founders of JCI Portugal, and I am now one of its senators and an international trainer.

JCI has various programs in the area of communication – international competitions in speech and debate – held all over the world. Recently, a world past president of this organization and I created a Public Speaking and Debating Academy, which we have held in various locations around the world under the umbrella of JCI.

For these reasons, JCI is a challenging organization that allows its members to experience intercultural communication, which is extremely important in this world that is growing more and more globalised.

Image 6.02. Official site of JCI Portugal (www.jci.cc)

Toastmasters

Toastmasters International is a non-profit organization focused on the art of communicating (speaking, listening and thinking) and in leadership training.

Toastmasters International serves almost 260 thousand members in 113 countries. There are more than 12,500 Toastmasters clubs in the world.

The mission of a Toastmasters club is to create a positive learning and mutual support environment, where each member has the opportunity to develop his communication and leadership abilities, which in turn increases self-confidence and personal growth.

Image 6.03. Official site of Toastmasters (www.toastmasters.org)

7

CHALLENGES AND SOLUTIONS

Communicators of the World, Unite!

Framework

As was the case with my previous book *Networking – Your Professional Survival Guide*, I decided to request contributions from people who have had significant communication experiences.

I launched the challenge through my social network contacts, and in this chapter I reproduce the communication stories sent by twelve professionals in nine countries, whose contributions I deeply appreciate.

They are situations that could occur to any one of us, and we can all learn from the solutions found.

Long distance intercultural communication

I worked at BP Oil, a business that operates in a global market, with operations in *x* number of countries with distinct cultures and languages. Communication is essential, and understanding how to communicate effectively with people who speak another language or who depend on different methods to reach a common objective is even more important. Now it is just as easy to work with people remotely as it is in person – intercultural communication is increasingly the new way of doing it.

The different cultural contexts create new communication challenges for the work place. Understanding cultural diversity is perhaps most important for people to understand that it is the key to effective intercultural communication, without needing to study individual cultures and languages in detail. We must all learn to communicate better with individuals/groups whose first language is not the same as ours.

Over the years, the challenges and opportunities in working and communicating with various colleagues from other countries and different cultures was a constant daily routine.

And it was precisely while working on these projects that *Communication*, N*etworking* and *Team Work* were crucial, I would say even indispensable, in reaching the proposed objectives, culminating with your «GO ALIVE». Although some of these projects were global at a BP Oil Group level, by its nature, the liberalization of the fuel market in Portugal was something new, and it was necessary for us to prepare well so that its implementation in our country would be successful.

Of course it was essential to know other markets with experience, already in a phase of maturity, and identify which of these would be the closest to our economic and fiscal reality. It was without doubt a great and gratifying challenge that from the start counted on the support and sharing of experiences, best practices and contingency plans from Spain, France, UK, Germany, Holland and Australia/New Zealand. Without that, such a feat would not have been possible to accomplish.

Communication was undoubtedly the key factor, accomplished through meetings and audio/video conference calls. They were, and continue to be, a constant throughout the whole process of completing each of the projects and of this one in particular. An effective communication strategy is extremely important, especially when it is done in a foreign language in a multicultural environment.

However, when working with people from a different culture, one must take into consideration pertinent aspects such as: tolerance, constructive participation, cohesion, effectiveness, efficiency, team productivity, awareness of time zone differences to keep all those involved aware, empathy, courtesy, curiosity and good will.

Lastly, it is just as important, or more so, to realize that the starting point of solving a problem is to assume that communication has failed; many problems would be resolved quickly that way. It is preferable to immediately assume the error than to persist in it.

Ana Mouronho de Almeida (Portugal)
http://pt.linkedin.com/in/ammalmeida

Use of collaborative platforms

The use of other methods to reinforce training, guarantee the sustainability of acquired knowledge and complete a follow-up is more and more a reality.

I refer to Webex and to teleconferences, which avoid participants having to travel and can be held in any location. As long as these meetings are adequately prepared, they become an effective means of reinforcing classroom training or giving a short instruction (no more than 2 hours).

This new approach requires that all those involved be knowledgeable about how it works. The facilitator must also be well prepared to manage such sessions.

Webex is a more complete method since it allows combining, beyond oral communication between participants, the simultaneous visualization of the presentation with the direct observation of the participants.

Ângela Pinto (Portugal)
http://pt.linkedin.com/pub/pinto-angela/a/7b2/665

When our nerves take over

I prepared for 3 weeks for my first seminar on investments and yet I still felt really nervous, when in reality I did not have any reason to feel that way. The seminar was directed at a few long-time friends.

For half a year, they tried to convince me to do this seminar, since they knew I had valuable knowledge of the subject and they were willing to listen everything I had to say. The situation was so good for me that even if I said something that was utter nonsense, they would still think I said something very clever – they trusted me that much.

Still, for me, doing this seminar was not something I was comfortable with. Every time I stood in front of people and had to say something, I felt like I became a block of ice and could not move and or say anything.

So, to overcome this, I started the seminar by sitting at the table with them and started talking. I talked a little bit and then gave them formulas for calculating the return on investment, for the first exercise.

After giving them the formulas (and still seated at the table), I wanted to write them on the board, so I had to stand up: «Oh, my God!» As soon I stood up I forgot everything I wanted to say or do! I could not help the participants with the exercise at all. In fact, when I tried to help them, I just confused them further.

Luckily, the people were in a good mood, which encouraged me to tell a small anecdote: «People who earn the most money are not always the best accountants» At this point, one participant simply said that they would just calculate the return by themselves, and then we would just move forward and learn new things.

It was amazing to see that people did not care about my mistakes at all, and didn't give any importance to the obvious lack of my presentation techniques. It encouraged me to think that I should just concentrate on what I wanted to teach my friends, in the best possible way, and not worry about unimportant things like my shaky voice, my face turning red, and my hands becoming ice cold.

In the eyes of my students I saw the desire to learn and their confidence in me, and that was much more important than how I did my presentation.

I can't remember if I was nervous during the rest of seminar. I only remember that several times I started the sentence and didn't know how to finish it. Once again, the participants helped me. I can't remember if I talked standing or sitting.

I was probably sitting most of the time, except when I stood up to write few words or draw a chart on the board. The seminar took 6 hours (with a small break for lunch) instead of the planned 2.5 hours (not very good planning, I know).

In the end, the participants and I felt great. In fact, I only showed 8 of the 20 slides I had, because people asked many questions related and unrelated to the slides I had prepared. So, I promised them I would do a second part to the seminar.

They are already asking when we can meet again, and proposing (just in case) maybe we should prepare for a two-day seminar instead of a one-day seminar, because we talk a lot and it takes more time than we plan. I am happy that my friends helped me to make my first seminar experience as pleasant as possible.

Aurelija Muzaite-Zavanelli (Lithuania)
http://lt.linkedin.com/pub/aurelija-muzaite-zavanelli/2/324/268

Being omnipresent

Communication Challenge: A Dutch client asked me in July to do an important 2-hour presentation on change management in the month of August in Rotterdam, at the same time I was planning on being on annual holiday in Sicily.

I would have had to travel 19 hours to arrive at their location so I proposed do a virtual experience using Skype. It was a delicate issue because it was debriefing the results of a company questionnaire regarding their own customer service satisfaction and the participants, themselves trainers, considered themselves to be experts in the matter. In addition, the session needed to be highly interactive.

Solution: We spent numerous days planning a 2-hour session in order to be sure that not only were the objectives met for the session, but that it was conducted in an interesting and interactive way to keep participants active and involved. In addition, we had to plan to manage all the various technical problems that could have occurred.

The session involved planning to divide all the participants (approximately 35) into small groups, which worked interactively and responded to questions we had designed.

We planned to have both an alter-ego facilitator « live» on site to manage the groups and a person who was assigned to document in the chat all the issues raised by the groups themselves. I presented myself to the group, my image projected onto one screen, and my slides projected simultaneously onto another screen.

Results Achieved: I saved the company the costs of travel and time lost, while achieving all of the session objectives, creating awareness among the participants regarding their need to improve customer service.

All this while demonstrating to the group that customer needs can be met in many different and innovative ways. Although it was a difficult audience, the evaluation on the session was very positive, and the client was equally satisfied.

Diane Fryman (Italy/USA)
www.linkedin.com/in/dfryman

Communicating on Facebook

There are many ways of communicating, and one of the most powerful is through the Internet, namely through social networks. Many people say that it is impossible to have real friendships merely using Facebook, since we have to know people in real life.

In my opinion, it is possible to create a good network of contacts and get new friends, new mentors via Facebook or social networks in general. The problem is that the people we want to meet and reach are also targeted by thousands of other people.

This is why you should try specific techniques of communication to convince them to be your friends or your mentors, even when they don't know you in real life.

If you want to connect with someone famous on Facebook or other social networks, start by analyzing their profiles on Linkedin, Twitter, personal website, YouTube channels, etc.

You should look for some details, such as: hobbies, common interests... In this way, you can build an image of your target. Why build such an image? The answer is simple: You can imagine the communication style of this person and what message to send to be able to build a relationship.

For me as a JCI Trainer, I was looking for foreign mentors and trainers. To that end, I did some research on the official JCI website, www.jci.cc, chose some names, sent Facebook invitations and they accepted me because my introductory message was simple, short and explained my objective.

Once they are on your friends list, your e-behavior and e-communication will be what convinces them to help you. I can give you a real example: in 2009 before the JCI World Congress in Tunisia, I was looking for sponsorship and could not find anything. Some sponsors were my Facebook friends, but I had never met them in person.

They noticed that my communication style was correct, preserving my personal branding and also promoting what I was doing as a trainer through videos, photos... etc. I spoke with many about sponsorship. Many others offered me money, without me asking for it, all because the conversations were done in the right way. Therefore I was able to attend the world congress for free.

These are, to me, the golden rules of communicating over social networks, in order to build relationships:
- We should be smart and simple.
- Communicate effectively with short messages.
- Keep a good personal branding and e-reputation.
- Choose the right people as targets.
- Present yourself quickly and efficiently.

Fares Ben Souilah (Tunisia)
www.faresbensouilah.com

And what if everything goes wrong?

In 2010, I was invited to give training in Japan. For this particular training, I needed hand outs for all participants to use during the training. The hand outs contained short stories and questions designed for the participants to reflect upon during the training.

Usually, the organizer takes care of printing such materials, but due to some communication mishaps, there were no hand outs ready for the training. Since I knew these were critical for the outcome of the training, I was devastated.

I first tried to find a place to print, with no luck. Then I asked for a projector and got one, only to find that it did not work with my computer. Time was now running out, and I was desperate.

People started to enter and prepare for the session, so I had to come up with a solution very fast. I decided that I would read the short stories aloud, so everyone in the room would hear them. I had to read every story twice, but in the end, everyone was very happy, and the training session was a success.

Kai Roer (Norway)
www.kairoer.com

Get people communicating

We aimed to improve the results and self confidence in the team. This applied to team members who did not know each other and to the ones who had worked together for more than 5 years. How to engage people in the audience into remarkable and lasting communication? First they need to be committed, showing their face and building trust, respect with a positive attitude in the room. But how?

The facilitator or trainer divided the people into groups, depending on how many people were in the room. The participant's task was to share about their achievement within the groups. This is the so-called «achievement talk». Each person had 2 minutes to deliver a talk on his/her achievement in life (personal, work, etc).

When everyone had presented their achievements, the group decided who the best presenter was. The trainer, who had all the nominations, called each of them to the stage, to share their achievement in front of everyone in the class. The best one is once again voted on by the entire crowd.

The outcome of this presentation activity in communication was:
- Positive atmosphere.
- Emotion.
- Recognition.
- Encouragement.
- Motivation.
- Relationship building.
- A better attitude for personal development.

Loreta Pi (Lithuania)
http://in.linkedin.com/in/ploreta

The element of surprise

The First National Meeting for L'Oréal Active Cosmetics, involving pharmacists at the national level, had a theme of anticipating the future, which needed, besides a global and involving presentation, something that could impact the session. At 9am on a Wednesday, 300 people were in an auditorium with the lights low and background music playing, ready to listen to yet another session where the invited speakers would speak on futuristic subjects and where the firm would present the brand strategies and new challenges for the following years.

The eruption of the Icelandic volcano had grounded the French speakers' flight, and so it was up to João Catalão and me to captivate and engage the pharmacists with a vision of the future.

My presentation, prepared and previously validated by the L'Oréal team, contained references to 10 global tendencies and 4 groups of consumers (Interactive Generation, Males, Females e 50+).

Everything was prepared to guide people into thinking about how they could incorporate these future visions into their business. We had found an interactive solution based on the importance of the intangibles and on the way of constructing histories that endure.

The end of the presentation was nearing, and the moment had come to test the concept. We projected a rock on the giant screen of the auditorium. A simple rock, without anything else. The idea was for everyone to think about what you could do with a rock.

I stated that I would like to sell it and asked who would like to buy a simple rock. Nobody did (as was expected). I explained the origin – it came from the Berlin Wall. A few showed interest (but the wall was made of cement!!). At last, I showed the true origin of the rock. It was lunar. The price rose, and so did the number of those interested.

In summary, a rock is just a rock. Unless it has a history and that story is relevant to whoever hears it. Based on that conclusion, assistants gave each person in the audience a rock identical to that shown, and I asked everyone to create their rock's story. In the case of doubt, and if somebody wanted to throw it at me, I had my defense ready – a welder's helmet and an industrial pot lid were my face guard and shield.

Simple ideas to create a smile in people. Managing smiles is increasingly the focus of all the stories we have in communication and in life.

Luís Rasquilha (Portugal)
www.linkedin.com/profile/edit?trk=hb_tab_pro_top

Overcoming cultural barriers

The first time I was invited to deliver training outside my country was in Cyprus.

Allow me to briefly introduce myself: I am Syrian, Muslim, wear a veil (hijab), shake hands with men but do not hug or kiss when greeting them and do not tell or share dirty jokes, though I do understand them, even if I pretend that I do not. So, I was in Cyprus.

I was lucky enough to have Filipe (the author of this book) as the head trainer and we were delivering a one-day seminar on effective presentation techniques.

Filipe, knowing me and knowing the organizers of this training, JCI Cyprus, sat down with me in a small orientation session to explain to me that the organization had some concerns about how to deal with me since they recognized cultural differences. Until that moment I had not thought of that as a challenge!

«So what? I am an open-minded and progressive person... I do accept and respect others» I was thinking to myself. But then Filipe decided that talking was not enough, he would prove what he is talking about by showing me a small surprise the organization had prepared for him in his room.

«OK! Why not?!» I answered Filipe and we went up to his room. All the way up I was thinking «Why didn't they prepare a surprise for me?!?!» At that moment I understood exactly what «shock» «surprise» and «challenge» mean!

The room was covered with condoms! Everything from the TV remote control to the door knob! The condoms had been washed on both sides! Filipe's welcome note said so! And I thanked God that they did not prepare a surprise for me!

On the morning of the next day, the training started and it was my turn to introduce myself. In my one-minute introduction I had to break the ice, or shall I say the wall, and put everyone at ease on how they should deal with me: The Different One! Because Filipe is just as crazy as they are or even more...

Here was my introduction:

«Well, I know that Filipe is the *Most Outstanding Trainer of the World*, and he is a good friend of yours, however, I was a bit sad when he told me that you prepared him a surprise and did not prepare one for me?! 'Till I saw Filipe's room».

At this moment, the faces of those who worked on the surprise turned red, and within a moment everyone in the room (who knew of the prank) burst into laughter. I paused and then I continued.

«Thank you for not surprising me!! Let's pretend that I am seven years old when it comes to jokes and everything will be just fine».

People laughed again and we spent an amazing time together with those fantastic and crazy people.

So, the next time you have a presentation and you think there is a gap between you and your audience, remember that a smile is a curve that makes everything straight. And that your biggest challenge can turn into your biggest opportunity by thinking positively.

P.S. If you are friends with Filipe on Facebook, you can see the pictures of his room;)

Mayss Al Zoubi (Syria)
www.linkedin.com/pub/mayss-al-zoubi-mayss-alzoubi-gmail-com/5/916/782

Dressed to communicate

When you deliver a presentation, one of the key elements as a presenter is comfort. Do you feel comfortable? First of all, you need to know your topic well.

After you are prepared to captivate your audience with memorable experiences, you need to feel good in your «shoes»; if you do not wear comfortable clothes, your presentation style can be affected.

I will give you an example: in one of my presentations, my shoes were hurting me. As a result, my voice, my posture and my speech flow were badly affected and I finished my presentation in a lot of pain.

Remember to wear comfortable clothes and shoes if you want to deliver a great presentation! Good luck!

Mihaela Stroe (Romania)
www.linkedin.com/in/mihaelalilianastroe

Communication has had a tremendous influence in my professional life. With more than 14 years in academic and professional education, being present in national and international venues has been a challenge that I seek to face with grit and determination.

Along this journey, the connection to the Toastmasters International and JCI (Junior Chamber International) families has been essential in reinforcing my levels of confidence when facing an audience. The learning and constant challenge to improve our communication techniques are important steps in evolving as speakers.

The new ways of communication through videoconferencing allow me to reach a global audience today. In 2010, on the occasion of the Global Social Media Day (www.smdayportugal.com), I was in Japan, England and Brazil, all on the same day!

Pedro Caramez (Portugal)
http://pt.linkedin.com/in/caramez

Going further

One of the most important skills as a trainer is to successfully communicate with different audiences. I think the quality of your training is related to how well you communicate. One of the keys to communicating with an audience is verbal communication. Therefore, it is important to choose the appropriate words, which cannot be the same all the time, depending on the specificities of each audience.

Throughout my 10 years of experience, I have seen that training higher positions, persons whose jobs require high standards of performance, allows you to use more elaborate expressions for easier understanding and opening their minds to a broader way of thinking.

Verbal communication

If the training is directed at people with a lower cultural level, you should use simple language, enabling full understanding. Thus, before the training, you should know and study about who is going to participate in your training. It can be the basis of good communication since it cannot be done in the same way in different audiences.

Surenchimeg Dulamsuren (Mongolia)
www.linkedin.com/pub/surenchimeg-dulamsuren/2b/a43/1a4

Bibliography

Booher, Dianna: *Communicate with Confidence*, McGraw-Hill, New York, USA, 1994.

Booher, Dianna: *Speak with Confidence*, McGraw-Hill, New York, USA, 2003.

Cardim, Luís: *A Comunicação*, IEFP, Lisboa, Portugal, 1992.

Carrera, Filipe: *Marketing Digital na Versão 2.0 – O Que Não Pode Ignorar*, Edições Sílabo, Lisboa, Portugal, 2009.

Carrera, Filipe: *Networking – Your Professional Survival Guide*, Edições Sílabo, Lisbon, Portugal, 2010.

Ericson, Jon M.: *The Debater's Guide*, Southern Illinois University Press, USA, 2003.

Hofstede, Geert: *Culturas e Organizações*, Edições Sílabo, Lisboa, Portugal, 2003.

Lewis S., Lynn: 10 *Steps to Successful Presentations*, ASTD, Alexandria, USA, 2008.

Molden, David: *Brilliant NLP*, Pearson, Harlow, Great Britain, 2008.

Navarro, Joe: *What Every Body Is Saying*, HarperCollins Publishers, New York, USA, 2008.

Oberstein, Sophie: *Beyond Free Coffee & Donuts*, ASTD, Alexandria, USA, 2003.

Pereira, Alexandre: *Como Apresentar em Público*, Edições Sílabo, Lisboa, Portugal, 2008.

Samuel Joseph, Arthur: *Vocal Power*, Vocal Awareness Institute, Ensino, USA, 2003.

Spielman, Sue: *The Web Conferencing Book*, Amacom, New York, USA, 2003.

Wilder, Cladyne: *Point Click & Wow*, Jossey-Bass/Pfeiffer, San Francisco, USA, 2002.

www.ingramcontent.com/pod-product-compliance
Lightning Source LLC
LaVergne TN
LVHW081353060426
835510LV00013B/1794